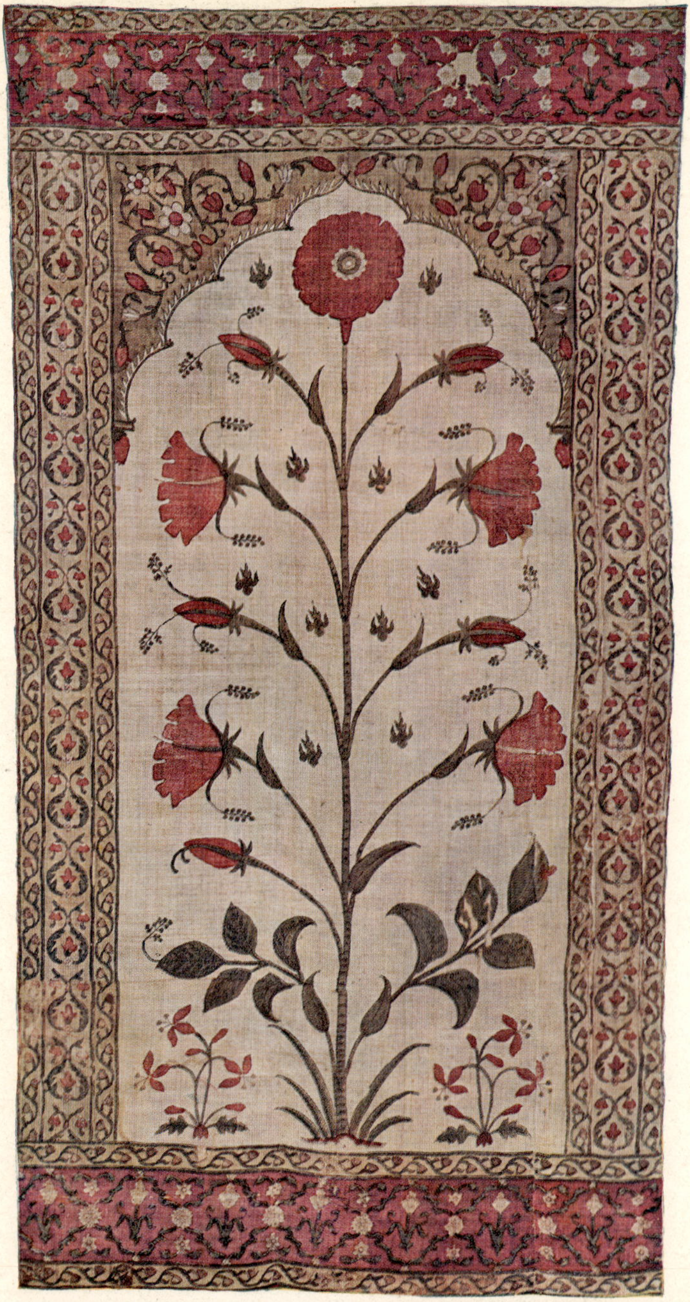

PLATE 1

EARLY SEVENTEENTH CENTURY PAINTED COTTON WALL
HANGING FROM RUINED CITY OF AMBER, SUG-
GESTING SYMBOLICAL COTTON PLANT.

BROOKLYN MUSEUM

THE HERITAGE of COTTON
The Fibre of Two Worlds And Many Ages

By

M. D. C. CRAWFORD

*Associate Editor of the "Daily News Record"
Former Research Associate in Textiles
American Museum of Natural History
Research Editor of "Women's Wear"*

G. P. PUTNAM'S SONS
NEW YORK : LONDON
The Knickerbocker Press
1931

Copyright, 1924
by
M. D. C. Crawford

All rights reserved. This book, or parts thereof, must
not be reproduced in any form without permission.

Made in the United States of America

The Constant Encouragement of

E. W. FAIRCHILD

OVER A PERIOD OF YEARS MADE POSSIBLE THE WRITING OF
THIS BOOK.

FOREWORD

This volume is a human record of a great fiber that has played a large part in the civilizations of two hemispheres and across more ages than modern civilization may safely span, and is still today the most important textile fiber. It is a history in paradoxes.

Cotton was ancient in India centuries before Cæsar conquered Britain. There was a trade in cotton between the Orient and Europe at least as early as the Crusades. Cotton fabrics were among the earliest objects of trade between the East and the West after Portugese da Gama opened the water route to India in 1497. Yet in the Eighteenth Century a half dozen British mechanics wrested the empire of cotton from the East within a vigorous generation of invention.

Columbus believed he had reached India because he found cotton in the Bahama Islands. The weaving and dyeing of cotton were well advanced in the New World many centuries before the Discovery, yet a single Yankee invention shifted the area of cotton cultivated in the New World within the territorial limits of the United States, where it was unknown at the time of the discovery, and cotton has played a great part since the first decade after the Revolution in the economic development of this country.

The early growth of New England is largely the record of cotton mill building. The early development of the South is the history of rapidly expanding cotton plantations and an international commerce in the fiber. Today the cotton mills of the South exceed in productivity the mills of the East.

What will the future of cotton be, what new shifts in plantation areas and mill concentration may we expect? Cotton was once the principal media of artistic expression. Cotton has become the great staple of necessity. Will it again have its golden age of loveliness, will some new fiber replace it in economic importance? Cotton has brought wealth and power, poverty and degradation in its history. What will its new relationship be to the social, artistic and economic history of America?

These problems are treated in a broad way in this volume, no particular phase unduly emphasized and all technical discussion, unless absolutely essential, avoided. At the same time the point of view of the historian, the ethnologist, the technician, the technical expert, the designer and the merchant have all been considered. It is a volume intended to be read not only by those actively concerned with specific problems, but those interested in the history of art and technology as expressed in fabrics.

INTRODUCTION

IT is my honest intention to keep this narrative as free from technical discussion and statistics as the subject will permit.

There is already a large and excellent body of technical literature for the mill engineer, and the tabulation of economic and industrial facts is easily obtainable and reasonably accurate.

It is my hope to present the story of the cotton fiber as a human drama—a drama that is by no means in its last act, but merely passing through one of its scenes.

Surely in a brief account of how the ingenuity of man met and conquered the many difficulties that surround the textile art, there is a deep and healthy interest. It will strengthen our modesty to compare our actual achievements of the last century and a half of mechanical effort with the distinguished accomplishment of cotton's golden yesterdays.

There are great traditions not alone in the accomplishment of loveliness in fabrics but in the spirit of workmanship and the underlying significance of effort, that are of incalculable value to us at this particular period.

If I may, therefore, through these pages induce men to look again upon cotton as one of the subtlest mediums of art; if in some measure I may direct the thoughts of manufacturers and laborers to a better understanding of the psychological value of interest in work, I shall be amply repaid for my efforts.

CONTENTS

	PAGE
FOREWORD	v
INTRODUCTION	vii
THE MARCH OF COTTON	xv

CHAPTER
I.—GENERAL REVIEW	3
II.—PRIMITIVE CULTURE	10
III.—PRIMITIVE TECHNIQUE	23
IV.—THE NEW WORLD	30
V.—PERU	46
VI.—INDIA	62
VII.—EUROPE	81
VIII.—ENGLAND	90
IX.—EIGHTEENTH CENTURY: THE AGE OF THE MACHINE	105
X.—COTTON IN THE COLONIES	125
XI.—THE MACHINE AGE IN THE UNITED STATES AND THE GROWTH OF THE COTTON PLANTATION	132

Contents

CHAPTER	PAGE
XII.—Mill Building in New England	144
XIII.—The South	163
XIV.—Research	178
XV.—Conclusion	213
Bibliography	233
Index	239

ILLUSTRATIONS

	FACING PAGE
PLATE 1.—EARLY SEVENTEENTH CENTURY PAINTED COTTON WALL HANGING FROM RUINED CITY OF AMBER, SUGGESTING SYMBOLICAL COTTON PLANT. (Brooklyn Museum)	*Frontispiece*
PLATE 2.—TEXTILE INFLUENCE ON OTHER TECHNIQUES	20
PLATE 3.—THE EVOLUTION OF SPINNING	24
PLATE 4.—PICTORIAL SURVEY OF THE TWO PRINCIPAL TYPES OF LOOMS	28
PLATE 5.—SPANISH INFLUENCE ON THE TEXTILE ARTS OF THE NEW WORLD	44
PLATE 6.—EVIDENCE OF COTTON CULTURE IN THE SOUTHWEST, MEXICO AND CENTRAL AMERICA	46
PLATE 7.—IMPLEMENTS AND PROCESSES OF THE TEXTILE ARTS IN PRE-INCA PERU	50
PLATE 8.—TYPES OF DECORATIVE FABRICS FROM THE GRAVE CLOTHS OF PRE-INCA PERU	56
PLATE 9.—CHARACTERISTIC DESIGN MOTIVES FROM PRE-INCA FABRICS	60
PLATE 10.—DISTRIBUTION OF THE WAX AND DYE, OR BATIK PROCESS	62
PLATE 11.—JAVANESE BATIKS, WALL PAINTINGS FROM AJANTA CAVES AND INDIAN PAINTED COTTONS SHOW SIMILAR ARTISTIC ORIGIN	76

Illustrations

	FACING PAGE
PLATE 12.—THE EARLIEST KNOWN COTTON WALL HANGINGS (17TH CENT.) PRESERVED IN GOOD CONDITION AND A FAMOUS ILLUSTRATION ON COTTON OF THE SIXTEENTH CENTURY	78
PLATE 13.—ANCIENT COTTON WALL HANGINGS OF INDIA AND FRAGMENTS OF RESIST DYEING FROM EGYPT	80
PLATE 14.—INFLUENCE OF INDIAN COTTON ON TEXTILE ARTS OF EUROPE	84
PLATE 15.—INFLUENCE ON ENGLISH CRAFT ARTS OF THE INTRODUCTION OF INDIAN CALICO	94
PLATE 16.—THE ERA OF MECHANICAL INVENTION	116
PLATE 17.—TEXTILE ARTS OF THE COLONISTS	126
PLATE 18.—ERA OF MECHANICAL ADAPTATION IN AMERICA	136
PLATE 19.—COTTON AND THE SOUTH	166
PLATE 20.—DEVELOPMENT OF COTTON YARN	180
PLATE 21.—MODERN RESEARCH IN DESIGN	200

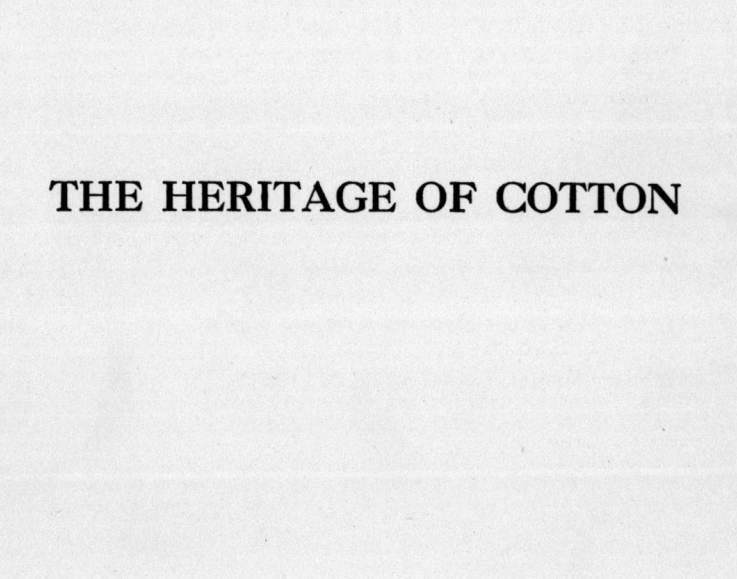

THE HERITAGE OF COTTON

THE MARCH OF COTTON

800–700 B.C.—Cotton cultivation and conversion are seen to have been long established in India, from references in the law books of Manu.

400–300 B.C.—Knowledge of cotton is first brought to the Greeks through Herodotus and the chroniclers of the campaigns of Alexander the Great in Central Asia and India. Other writings of the period refer to the exportation of Indian cotton products through the carrying trade of the Arabs.

300–200 B.C.—Cotton cultivation and conversion reach the shores of the Mediterranean via Asia Minor.

70 B.C.—The Romans use cotton tents, awnings and canopies. Compared by Lucretius with the white clouds of heaven.

70 A.D.—Pliny reports cotton cultivation and manufacture in Upper Egypt. The priests' garments are made of cotton.

100–200 A.D.—Cotton is grown in Elis and there also manufactured into hair-nets. This is the first recorded instance of cotton grown and manufactured on European soil, but the industry remained isolated.

200–300 A.D.—Arrian writes that calicoes and muslins are shipped from India to Adule, an Arab port on the Red Sea.

600–700 A.D.—Cotton reported cultivated in China as a decorative plant.

798 A.D.—Cotton first reaches Japan through a shipwrecked inhabitant of India. The cultivation was later abandoned.

The March of Cotton

912–961 A.D.—Cotton culture and manufacture are firmly established in Spain under Abdurrhamans III, and also in Sicily under Arab rule.

1050 A.D.—This is the date of the earliest extant specimen of cotton paper, in the manufacture of which the Arabs of Spain are said to have excelled.

1096–1270 A.D.—The Crusades introduce Europeans to the varieties of Levantine and Occidental cottons, disseminate a knowledge of cotton goods, and initiate first an industry in the Crusader states of Asia Minor, and later a lively trade in cotton goods between the Italian city states and Asia.

1200–1300 A.D.—The Tartars introduce cotton cultivation and use into China. This knowledge is afterwards introduced into Korea.

1200–1300 A.D.—The earliest references to cotton appear in contemporary French and English writings. Cotton was first used for candle-wicks in England, and also as trimming for doublets. In France, cotton seems to have been used to make hats. Other references of the same period mention cotton as used in the form of a defensive pad in warfare and also as part of fortifications.

1200–1300 A.D.—Barcelona flourishes as a cotton manufacturing center, specializing in cotton sail cloth and fustians.

1320 A.D.—Oppel claims that Ulm in Germany is the first place in Central and Northern Europe where cotton is spun and woven. Venice claims the honor for Europe.

1350–1400 A.D.—Cotton cultivation reaches the Balkan peninsula through the invasion of the Ottoman Turk.

1375 A.D.—English literary references indicate that cotton goods were being imported as a usual thing.

1492 A.D.—Columbus discovers cotton in the Bahamas. On his return trip, Europe gets its first glimpse of Sea Island Cotton.

1520 A.D.ca.—Magellan reports cotton in Brazil.

The March of Cotton

1560 A.D.—Ghent and Bruges are famous for their printed cotton goods.

1592 A.D.—The Portuguese reintroduce cotton into Japan.

1600 A.D.—This is the date given by some authorities as the beginning of real cotton manufacture in England, coincident with the coming of Flemish refugees from the Netherlands.

1619 A.D.—Cotton is grown by the colonists along the rivers of Virginia.

1619 A.D.—The first negro slaves are imported into the New World.

1621 A.D.—London wool merchants protest against the growth of cotton manufacture, alleging that 40,000 pieces of mixed cotton and linen fabric are being produced yearly in England.

1641 A.D.—This is the date set by George Bigwood as the real beginning of the cotton industry in England. Prior to this date, he says, cotton was only used in England to make candle wicks.

1678 A.D.—Pamphlets indicate that cotton goods are gaining popularity in England.

1700 A.D.—Cotton cultivation in North Carolina furnishes one-fifth of the population with clothing, the cotton being mixed with other fibers to produce cloth. Every farmstead has its cotton patch.

1700 A.D.—First law in England forbidding the use of cotton in the interests of wool growers.

1700–1793 A.D.—The West Indies and Brazil are the great cotton producing countries of the New World.

1721 A.D.—Parliament passes a second law protecting the wool interests in England, fining any one who wears a dyed or printed calico.

1733 A.D.—Kaye invents the flying-shuttle.

1736 A.D.—The Manchester Act is passed, allowing cotton and linen mixed calico to be manufactured, while importation of Indian goods is still forbidden, thus giving Lancashire the monopoly in cotton goods.

The March of Cotton

1750 A.D.—30,000 people in Manchester and Bolton districts are concerned with cotton manufacture.
1753 A.D.—South Carolina sends a few pounds of cotton to London.
1764 A.D.—Hargreaves invents the Spinning Jenny.
1764 A.D.—Eight bags of cotton are sent from Carolina to Liverpool.
1766 A.D.—Manchester and Bolton manufacture 600,000 pounds sterling worth of cotton and linen goods per year.
1769 A.D.—Arkwright patents the spinning frame.
1770 A.D.—Three bales of cotton go from New York to Liverpool, ten from Charleston, four from Virginia, and three barrels from North Carolina. (Note: A bale or bag at that time was computed at 200 lbs.)
1775–1783 A.D.—Cotton manufactures in America are stimulated by the cotton goods famine incident on eight years of war.
1779 A.D.—Crompton invents his Spinning-Mule.
1783 A.D.—The first piece of cotton goods entirely made of cotton is produced in Lancashire.
1785 A.D.—Cartwright invents the power loom.
1786 A.D.—600 pounds of American cotton are shipped to Liverpool.
1787–88 A.D.—The first permament cotton factory built of brick, at Beverly, Massachusetts, is put into operation by a group of men headed by John Cabot and Joshua Fisher. It was not an economic success.
1788 A.D.—A factory is built at Philadelphia equipped with expensive machinery for carding and spinning cotton.
1788 A.D.—Richard Leake of Savannah announces a new staple and decides to experiment with eight acres planted with cotton seed.
1789 A.D.—127,500 pounds of American cotton exported.
1790 A.D.—Samuel Slater migrates from England, and puts up a factory at Pawtucket, Rhode Island, embodying the coveted English inventions.

The March of Cotton

1790 A.D.—3,138 bales of 500 pounds each are produced in America, and 379 bales are exported.

1793 A.D.—Eli Whitney invents the cotton gin.

1812–1815 A.D.—War with England stimulates American manufacture.

1815 A.D.—Boston Manufacturing Company founded with power looms.

1820 A.D.—Factory system begins to be applied in England to the weaving industry as well as spinning. At that time there were seventeen times as many hand looms as steam looms in the country.

1832 A.D.—Invention of ring spinning.

The Heritage of Cotton

The Heritage of Cotton

CHAPTER I

GENERAL REVIEW

THE early history of cotton lies in Asia and in the New World, in ages only partially historic and among peoples wholly alien.

Our first literary record of cotton is in the vague phrases of a dead language. The most ancient cotton fabrics are the remains of a civilization that matured and vanished in the New World while Europe was still a barbarous wilderness. As this delicate seed hair first appears in the traditions of Asia, or the marvelous grave cloths of pre-Inca Peru, it is already a finished achievement, complex and varied in technique, highly developed in æsthetic values, the fruit of long ages of development and accomplishment, its standards beyond our latest skill.

In Europe cotton was known but was of little importance in commerce until the hardy mariners of the Sixteenth Century linked the Orient and the Western Hemisphere to Europe. It had no great industrial significance, until the mechanical genius of a few inventors in England, and one in the infant republic of the West gave to English speaking people that control of cotton we still enjoy.

It is today one of our chief forms of wealth, quoted on every bourse of the world, a great agricultural staple, a great industrial factor. It had as well its ages of beauty, still preserved for us in priceless webs of living color. Beyond these lie the misty centuries and vague races who first discovered cotton in nature, developed the tools and implements, methods and processes that underlie both the technique and arts of today. All that our own age has done, is to make automatic and mechanical movements and principles that were ancient when life was new along the Ganges, when the ancient civilizations preserved beneath the sands of coastal Peru were at their earliest dawning.

With reasonable accuracy we can trace each phase of cotton history as the fiber affects Europe and our own Colonial and early national life. It is true, that this record is neither extensive nor wholly clear, until modern times. But still each phase is sufficiently definite and never lacking in interest.

Spain was the first nation of modern Europe to know cotton both as an agricultural product and as a textile fiber. The Moors introduced cotton into Spain in the Ninth and Tenth Centuries and achieved great skill and artistry in its conversion.

What part cotton played in the little trickle of Oriental commerce which began after the first Crusade, it is difficult to say. There is uncertainty in the names of fabrics and the lack of scientific knowledge as to the character of fiber that makes this question difficult to determine. Cotton and cotton cloths may easily have passed unnoticed in the commerce of these and even earlier times.

In England we first hear of cotton in the late Twelfth Century as candle wicks, embroidery yarns, and as a vegetable wool from the Levant, to be mixed with flax, or sheep's wool, in the heavier, cheaper fabrics of the poorer classes.

No doubt, the traveled scholars of the Monasteries knew its true character, yet the common belief lay in a series of quaint legends of which the delightfully unveracious Sir John de Mandeville was, if not the father, at least an earnest supporter.

Cotton was supposed to be the wool of certain mysterious Scythian sheep. These lambs grew on shrubs, each cradled in its downy pod. Except for the fact that the stalk was attached to the soil, they were like the little downy creatures who gamboled in the English fields. Fortunately this stem was flexible and permitted them to bend down and graze on the adjacent herbage. When, however, all grass within this narrow orbit had been eaten, the lambs naturally and wisely proceeded to expire. Both wool and flesh were then available. Short of a fire breathing dragon, no animal could have been more satisfactory to the Middle Ages. The actual cotton plant was tame beside it. To whom belongs the distinguished honor of this discovery, I can not determine.

It is difficult to understand why Italy, more open in many ways to the influence of the Near East than Spain itself, did not acquire some skill in cotton spinning and weaving. Her artisans drew a rich inspiration from the eastern arts and had learned how to weave silk and eventually how to raise the delicate, little moths, and supply her own raw material. France, Flanders, and even England acquired in time some degree of Italian skill in this apparently equally difficult medium of expression. But cotton eluded these masters of the loom for centuries. It may have been that in cotton weaving, spinning and dyeing there were technical difficulties a little beyond their power. At least, we know that when Europe a few centuries later, with greater skill and knowledge, attempted to produce all cotton fabrics in competition with India, they met with failure, until the Machine Age.

There is a rather scanty but none the less interesting classical history of cotton. The Greek historians before the Christian Era knew of the fiber and something of the methods of decoration. Alexander the Great carried back from India cotton cultivators and craftsmen and settled these in Asia Minor. There is even some evidence that the arts spread temporarily to the Greek mainland.

The Chinese had evidently heard some rumors regarding the fabulous Scythian lambs long before the legend was current in Europe, but the commercial explorers of Old Cathay were either not gifted with the imagination of the Moyen Age, or were not encouraged by their masters to employ their talents. Hence, they merely quote as heresay vague reports regarding "water sheep" which sound suggestive.

Sir M. Aurel Stein, the explorer of the ruined City of Turfan on the Gobi Desert, mentions the finding of cotton cloth and dates these fabrics from the Second to the Fifth Centuries of the Christian Era. This mysterious ghost of a city, was once a busy market place on the old caravan route between the Near East and China. If the distinguished scholar is correct in his fiber analysis, it is proof that cotton fabrics were articles of trade between India and China in very remote times.

There are certain stories that the beautiful blossoms induced Chinese florists to cultivate the plant as a garden flower, but Marco Polo, in the Thirteenth Century and later Arabic merchants, state positively that cotton was only known in the southern province of Fokien and was then a recent introduction from India.

Egypt in very remote times maintained an extensive commerce with both India and indirectly with China, but neither silk nor cotton are found in the Egyptian tombs until late in the Sassanian Empire which begins in the Eighth Century A.D. This does not prove, of course, that Egyptian merchants and travelers did not know of cotton or silk; it merely proves that other things were regarded as more important in commerce.

Cotton culture and the arts of cotton conversion, more particularly the distinctly Indian craft of resist dyeing, spread to the islands of the Indian Ocean at a very early date. Java is supposed to have received cotton between the Third and Fifth Centuries through the efforts of the Buddhist monks and traders. It is even believed to have reached Japan a little later, but if so, it left little if any trace. The fabric arts of later Japan, built so largely upon the cotton arts of India, were the results of the efforts of Dutch traders bringing the craft secrets from India in the Seventeenth and Eighteenth Centuries.

It is not difficult, perhaps, to understand why cotton spread from India to the Islands of the Indian Ocean and not into China or Egypt. All people at a certain stage of culture are good weavers and spinners, their crafts well established and adjusted to local supplies of raw material and domestic needs. The introduction of a new fiber or fabric or craft is always attended with difficulty, unless the advantage is very great and apparent. The arts of cotton only spread when accompanied by a migration of Indian craftsmen, friendly climatic conditions and where the cotton fabric itself was wholly suitable.

All of this evidence points conclusively to Southern India as the source of origin for cotton so far as Asia, Europe, and Africa are concerned. The discovery of cotton in the New World, its prevalence and high development all through Spanish America, came, therefore, as a great surprise. Even at this late date we have not solved this great enigma. We have learned, however, that the cultivation of cotton in the Americas, together with a technical development, at least as complex as Asia's, existed here prior to the earliest certain date we can give to cotton in Asia.

In a later chapter I will discuss the influence upon the social habits of Europe, particularly England, due to the introduction of cotton fabrics from India during the Sixteenth and Seventeenth Centuries. This was followed almost immediately by the discovery in the Western World of an almost inexhaustible supply of raw material as a further incentive to artistic and industrial development.

If there is one thing more certain than another in the history of cotton, it is that its beginnings in both technique and art retreat beyond the veil of history into the earlier phases of human development.

The modern mill, with its minute divisions of processes, its intricate and rapidly moving machinery, is incomprehensible to the most intelligent lay observer. The so-called hand implements and processes which preceded the machine, while perhaps a little clearer because of the absence of equipment, to control and distribute power, are only a degree less confusing. It seems wiser, then, to first firmly establish the fundamental ideas regarding spinning and weaving, to rationally develop these implements and methods as these mature in human consciousness, than to explain such intricate machines as those now in use or their more direct antecedents.

We must remember that cotton, as wool, flax and silk, was first discovered by people at a very low stage of culture. It was one of the many fibers tried that met successfully the varying needs of human life over an immense period of years, and has survived across ages untold and largely unknown.

How did man first learn to spin, to weave, to dye, to create design in its simplest form? What was the motivation, the guiding impulse? As the records and training of childhood in a large measure explain the adult, so in the infancy of the race, are lessons of value to the present age.

Each civilization has the happy and wholly satisfactory habit of regarding itself as the culminating epoch of all culture. It is true that each period contains some traces of all the achievements and visions, hopes, defeats and aspirations of the past; but there has been loss as well as gain, nor has the upward climb always been direct. There are many things worthy of retrieving even in our most remote past. No great fundamental history of any art may be truly understood if studied merely from the vantage points of a single epoch.

CHAPTER II

PRIMITIVE CULTURE

IT is natural that the entire problem of textiles should be confused with the single modern phase of cloth making. The woven web plays so important a part in modern life, we forget that in primitive cultures it does not appear until a comparatively late period. Spinning, weaving and dyeing in the Dawn Ages are independent arts which served the needs of society separately for an infinitely longer period than when combined.

My purpose in this chapter is to correlate in a brief sketch the basic principles of the textile arts and to show these in their simplest, most elemental forms. To outline the primitive stages of evolution up to the time when loom and spindle, the first crude assays in design, color and the deliberate production of cloth, bring the story within the historic limit of our own hemisphere and Asia.

Spinning, the act of combining two or more comparatively weak filatal elements, through twisting into a comparatively strong yarn, cord or thread, precedes by a full cultural cycle, the crudest idea of weaving. We have still to find any civilization archaic or modern, so elemental as not to be proficient in the arts of spinning. Almost every conceivable material, both animal and vegetable, have been used by primitive peoples for cordage. The list is endless. Our earliest ancestors evidently conducted a constant search in nature to find materials to answer their varying filatal needs. There are tribes today in inaccessible jungles, who have no knowledge of weaving in any of its forms, who neither make baskets nor pottery, nor practise the most rudimentary agriculture, but who still know how to spin and have apparently always possessed this knowledge. Even among the few races who can not produce fire by artificial means, the art of spinning is a familiar practice.

It is safe, therefore, to reckon cord making as among the first technical achievements of men. Still there must have been a time when man did not know how to spin.

There is no doubt that early man developed in continental areas and was a companion of dangerous, carniverous beasts, and faced these dangers with only his native shrewdness and a certain swift strength as his protection. Strong as were his arms, they were no match for the bears; terrible as was the grip of his blunt fingers, it was as nothing to the rending claws of the tiger. If he could not devise some weapon that would equalize these odds, he must have perished in the contest.

From immemorial silt, from cave débris, from river drift and marl pit, wherever in our dawning man made his lair home, we find jagged stones, rounded on one side, small enough to cup in the hollow of the hand, yet large and heavy and sharp enough to be a dangerous weapon at close quarters. We call these handstones, the first tool weapon of the human race. To conceive even so elemental a weapon, man had to break up pieces of flint and use judgment in the selection of the most suitable fragments. No doubt with the aid of this stone he fashioned a club of gnarled wood. This was a reasonable sequence of ideas, since the act of striking, throwing and stabbing are natural movements.

At some time there must have come a man, to whom the jagged handstone and the wooden club were insufficient. He did not accept the universe as he found it. In other words he thought. He learned to make this clear distinction between the club and the handstone. With the club he could strike a shrewd blow, beyond the reach of danger and this sufficed for the lesser beasts. With the stone he could strike a downright blow, which even the greater beasts respected. Yet to strike this blow he must come perilously within the reach of claws and fangs.

If a way might be devised to combine the comparative security of the club with the force and impact of the stone, it were a great matter. But stone is stone, and wood is wood, and excellent though each be in itself, according to its nature, they will not grow together. Some hint he may have received from the intertwining creeper, the tough lianas coiled like serpents about the boughs of trees, or the long sinews in the legs of the deer and the auroch might have given him his first idea. No question he has dragged these out again and again to study mutely their peculiar nature, to experiment with their slender toughness. In the end he loops these sinews about the stone and attempts to fasten it to the end of the club. He learns that if he doubles these sinews, they are better than single, but if he not only doubles them, but twists them, this double and twisted element, is not only still stronger, but the spiral character of the twists forms a friction producing surface, which prevents the binding from slipping.

This twisting of sinews together is spinning. Spinning indeed, in its crudest form, but still true spinning. He has combined with the aid of his strong fingers two comparatively weak filatal elements and fashioned of them one comparatively strong cord. And this suffices for his purpose. Stone hammers will be fashioned to wooden handles by this means for more thousands of years than there are centuries in modern history.[1]

This twisted cord of sinew marks a great change in human life. New uses for cordage quickly developed. For example, man had noticed that flint broke in razor-edge flakes. It would not be long, therefore, before he would fasten with twisted sinews one of these sharp edged flakes to the end of a slender sapling. He would then have a stabbing stick or spear capable of piercing the toughest hide. This would rapidly develop into a Javelin or throwing spear. Fish lines, snares, in some parts of the world the sling, and nets, would be made from twisted sinews or tough grasses and shredded barks at about this era of culture.

The idea of cordage of any kind is so universal, that it might well have had multiple origins, the idea occurring to many men in many parts of the world, separated by great reaches of time.

The next use to which cordage was applied is so unique, so original, that it could have but one single source of origin.

[1] I am well aware that stone axes and points of stone have been secured in holes bored in wooden hafts. Whether this method preceded the binding of sinews or not, makes little difference in this narrative, for I am describing the primitive development of textiles, and not the serial culture of man. My hope is to drive home in a few examples the importance of textiles before cloth making.

This second man in the great primitive trilogy of inventors must have been gifted with analytical reasoning powers, been able to carry his deductions through several phases of mental experiment before he actually put his ideas in operation. In other words the final results of his thought was highly composite.

This man sees in cord and wood and point of flint new and undreamed of possibilities. To his mind flint is the most perfect material, nor can he conceive of a stronger, tougher cord than that made of twisted sinews. He, therefore, turns his investigation towards the properties of wood. He is familiar with many kinds and each has its separate function in his life.

Chiefly, however, his interest centres about the smooth barked, straight saplings which are best suited for haft of stone hammer or shaft for flint tipped spear. In making new weapons he has often marveled how the strength of the sapling grew as he put forth his own strength. The more he bent a sapling the greater grew its power to resist, and if it escaped occasionally from his firm grasp in its upward swing, more often than not it gave him a shrewd blow. This was an indication perhaps of displeasure on the part of the wood, and yet once the wood had been subdued, this spirit of discontent left it and it became his willing servant and stout friend.

In time it occurred to him that a sapling might be more easily severed with the jagged stone edge if one end were tied down. Here he discovers that it is not only a convenience but a miracle, for the strength he had marveled at in the bent sapling and the pliant toughness of the cord, when combined, froze into a silent struggle between cord and wood.

He runs his calloused hand along the straining curve

of the bent sapling, it is not unlike his own flexed muscle, when he puts forth his might. Yet it is still the familiar sapling of his many experiences. In wonder he looks at the cord he himself has fashioned from the sinews of the auroch. It is just the same cord he has known for many friendly days and yet how different! Cord and wood have suddenly grown alive in this new relationship.

Like a child he touches the straining cord and it answers with a deep, rich note of music. It is the spirit of the wood making protest at being bound! None the less he touches it again with the same result; and yet that note, deep as it was, strong as it was, was not the tone of anger. It is like the call of his mate, like the song that springs to his own hairy lips when the long days come again with the flowers, and the fish leap in the shallows. He strikes it tentatively with his flint hammer and the cord almost snatches it from his hand. He pushes against the cord with the blunt end of his spear, the spear springs away like a living thing, and the cord hums with satisfaction!

What are this man's processes of thought. How does he arrive from one experiment safely to the next, from dream to idea, from idea to action, from action to accomplishment? No man can say. The night of our long climb is lit by the brilliance of a few great intellects, nor is there any rule to measure either their mental processes or the sequence of their appearance. Among the truly great inventors of all the ages, I place the Bow Maker!

In all eternity there must have been the twilight instant when this miracle was still but the arched sapling and the singing cord; and that instant when the idea flamed like a comet in the midnight firmament and as clear as his own image in a crystal pool, he sees how all things, strength of cord toughness of wood, power to throw objects aside that touch the cord, all mean, have always meant, must always mean one thing, the Bow!

He has bent his last sapling, and attached to either end the twisted cord. In moist, triumphant hand he holds the fruit of his thought and listens to the music of the throbbing cord. It will throw a stone further and surer than his sinewy arm. This is good, but in his mind there is a still greater thought. For if he, himself, the creator of the bow may throw a spear, why may not this creature of his mind also throw its spear? (He has not yet learned to call them arrows). He selects a straight stick, scarce larger than his finger and yet long enough to span the arch of the bow. He tips it with a flake of flint and rubs a notch in the end. Timidly he draws his first arrow, releases it and follows its flight with dazzled eyes. Again and yet again he performs this miracle with increasing satisfaction. Now he fashions a bow just to the verge of his might to draw and uses great care in selecting arrow shafts, in balancing the tips. And later, either he or some of his descendants fasten to the butt the guiding feathers of the fenney goose.

Now comes the never-to-be-forgotten hour, when he makes first assay of his new powers against his ancient enemies. He sees the arrow redden in the heart's blood of the snarling wolf, he sees the gray feathers encrimsoned against the tawny shoulder of the tiger-devil. His hour has come at last, and he walks upright. No longer shall the lurking shadows of the forest trail drive him chattering in fear to the protecting tree tops. His is the power, safe himself, to send from afar a lethal message. Men will come and go for countless ages and the castellated ice will cover completely this verdant forest, where now he roams in the sweet security of mastership, nor will men ever devise a more perfect, more certain weapon till the swing of the cycles bring gun powder and the rifled barrel.

And all these—axe, snare, sling and bow—are in a sense results of the twisted sinews of the first Cord Maker!

The first weaver was he who made the first fish weir. He had observed that it was a little easier to secure his prey in parts of the stream where a wind-blown tree formed an obstruction which permitted the fish to escape only in one direction. Obviously the gods of the forest sent these fortuitously fallen trees in answer to the offerings of the tribe. Still the gods were notably poor fishermen and did not send either enough trees or trees in the right place to suit the growing demand of the ravenous appetites.

This man dares to imitate the gods to improve indeed upon their careless methods. He selects a shallow ripple, between two deep pools, drives in upright sticks in a loop to suit the vagaries of the water. Between these, he intertwines saplings of pliable vines, so that the water may escape but the fish be retained. All that he thought he was doing, all that he hoped to accomplish and all so far as he ever knew he had accom-

NOTE: "The missile bow, whatever its form, I regard as a comparatively late development in culture, preceded by the throwing stick and sling, and, in my opinion, a probable development from the throwing stick and nowhere to be regarded a direct invention in any of its existing forms. From studies made with the late Frank Hamilton Cushing I consider all missile bows to be genetically related and as having a common, rather than several, independent origins. They all, at least in my opinion, have an identical morphology."—Stewart Culin, Brooklyn Institute Museum, Brooklyn, New York.

plished, was to build a fish trap in imitation of wind-blown trees. As a matter of fact he is the father of weavers and the latest, most sumptuous fabric of our times, is covered by the same generic definition as this rude texture of upright staffs and intertwining withes.

For weaving is the act of interlacing at right angles two filatal elements in such a way that friction holds them in one compact entity.

Not long after this great invention of the fish weir, the growing wealth of the tribe in food, gave individuals leisure for further experiment and mats of rushes were intertwined in imitation of the fish weir and wattled huts built originally above a spearing platform as protection from the wind and rain, began to make their appearance. Before long came that period of culture when man left his caves and tree top homes and lived in spiled villages above the shallow waters of lakes. Here he was reasonably secure from danger of attack from marauding animals and devastating forest fires.

The task of fishing was relegated to the women. And women not only repaired but built the fish weirs. So all women came to a certain skill and understanding of the arts of interlacing two sets of sticks into a strong open texture. In other words, became weavers.

Woman's life was changed by the weapon-ingenuity of man from one of nervous dread to comparative security and comfort. No longer does she clutch her latest born and so perilously and preciously handicapped flee when the tiger slips like a gray flame into the clearing. Now he may roar his loudest, tantalized by the delicious scents wafted across the water to his quivering nose. If he shows his evil, wrinkled face at the bridge-tree, the men will fill him full of arrows and there will be a great feast that evening, and the chief will have a fine new tiger skin to sit upon before the council fire. No longer does she huddle in mute agony in the corner of the cave, with her little, whimpering bundle of tenderness, while the gaunt silent men return day by day fishless from the pitiless river. Now plenty warms her life and there is still to eat and eat again. Her man, lean-hipped, burly shouldered, with corded, hairy arms and soft anxious eyes; a bow of seasoned wood, a flint tipped spear, an edged stone cunningly hafted in a stout sapling, the keen delight of the hunt, the riotous joy of the evening feast and the satisfaction of talk with his peers about the council fire; for him these suffice. But not for her. She demands some outlet for the pent up energies, some direction for the creative instincts fostered by the new relationship she has developed towards this particular tiny wattled hut, carpeted with rush mats and the skins of animals.

So one evening when her man returns from the hunt, he is shown the first crude basket. For the woman in her new life has always something to carry. First the fish must be transported from the weir, then there are the swift, timid forays into the adjoining forest to gather nuts, and fruits, berries, edible roots and grasses, which have become an important part of the daily menu. The basket is, perhaps, the first device created by human ingenuity that may safely be called a luxury. It is weaving, a direct result of the fish weir and mattings. The man looks at it with superior toleration. It is obviously neither weapon nor trap, nor can one make upon it interesting and satisfactory noises. It seems, however, upon reflection to be an overturned hut, a sort of propitiatory offering to the gods of the winds to dissuade them from blowing down the real hut. Such a ritualistic theory is sound, for in spite of the gods' power for mischief, they are notably easy to deceive.

That night he showed it at the council fire of the elders. They turned it over, peered into it, smelled it, and wisely shook their heads. Woman's work of small moment! None the less he carried home in it that night the arrow points he had fashioned while listening to a learned discussion around the fire regarding the habits of the Wolf "who eats the sun each night."

The men, through better organization and improved weapons, have driven the more dangerous animals from the little clearing on the shore and the woman finds in the forests more and more things of value. Roots, grasses, fruits and berries, she can gather with more or less impunity and for these she needs more and different types of basketry. She finds as well that certain of these foods, can be dried and stored and this is a further occasion for weaving baskets, and hence increases her quest for fibers, that may be spun and woven.

To all of these forces there comes the added incentive of color, the desire for chromatic sensation. Heretofore fear, hunger, desire for security and rude comfort have entirely governed the quest. Now comes the first, great æsthetic impulse, latent always, awaiting this hour to blossom into centuries of beauty.

No one studying the arts of primitive people can fail to be impressed by their passionate love for color and design. Almost every implement, every object capable of being designed and stained is decorated.

Primitive peoples have a keener if narrower vision of nature than civilized man. They see animals, fishes and reptiles, as well as vegetable matter in all varying forms and can identify the objects of their quest by the

PRIMITIVE CULTURE

The impulse towards ornament is very ancient in human culture. Many peoples who are not cloth makers, produce design subsequently used in cloth and basketry. Certain types of design originated in body painting and tattooing. Many techniques, first developed in basketry, later appear in textiles and are subsequently transferred to pottery.

1—Prehistoric lost color ware or resist dye pottery from Central America.
American Museum of Natural History. *(Page 59)*

2a—Textile pattern on prehistoric pottery from the Southwest. *(Page 40)*
American Museum of Natural History.

2b—Textile pattern on prehistoric pottery from the Southwest. *(Page 40)*
American Museum of Natural History.

3—San Carlos basketry patterns produced by non-cloth making tribe.
American Museum of Natural History. *(Page 19)*

4—Rug of fur applique from Koryak Tribe of Siberia. *(Page 20)*
American Museum of Natural History.

5—Valiente Indian knitted bag from Central America. *(Page 20)*
American Museum of Natural History.

6—Valiente Indian knitted bag from Central America. *(Page 20)*
American Museum of Natural History.

7—Pima baskets showing weaving pattern by non-cloth making people.
American Museum of Natural History. *(Page 19)*

8a—Etched bark belts from New Guinea. *(Page 20)*
American Museum of Natural History.

8b—Etched bark belts from New Guinea. *(Page 20)*
American Museum of Natural History.

9—Detail of decoration, interior of chief's house in New Guinea. *(Page 20)*
American Museum of Natural History.

10—Tattooed Marquesas Islander. *(Page 21)*
American Museum of Natural History.

11—Samoan fabric of pounded bark known as Tapa, showing textile design produced by non-weaving people. *(Page 36)*
American Museum of Natural History.

12—Samoan implements used to pound paper mulberry bark and print patterns. *(Page 36)*
American Museum of Natural History.

13—Samoan tapa cloth. *(Page 36)*
American Museum of Natural History

PLATE 2

least shade of color visioned. They acquire fine distinctions in color senses, since it is necessary to distinguish the tawny shade of the lion's mane from the dun color of the deer, to catch the glint of the fishes' scales in the deepest pool, to distinguish at a glance the edible from the poisonous mollusk.

The marked preference of all savage people for different tones of red, which still survives in our own consciousness, is because this color has always pleasant associations, whether it be in fire, the warm blood of the slaughtered animals, the tones of ripened fruit, or the clay which protects from annoying swarms of insects.

But there is obviously a vast intellectual difference between seeing color in nature and reacting to color, and in deliberately using color to produce effects. The first coloring was, no doubt, accidental. The juices of fruit, the liquor of certain shell fishes and roots and edible barks, left upon the glistening skin visible traces and between these colors and the satisfaction of eating there was a definite mental reaction.

So gradually, the art of body painting and staining developed. I need only cite a few of the more familiar instances to establish this first of all arts in its proper status.

The little peoples who lived beyond the walls in northern Britain, were called by the Romans, Picks, the Painted People, because of their habit of tattooing their bodies. The Sioux warrior, the Zuni priest, the African spearman, even the Australian bushman, (lowest in human culture), decorate themselves in marvelous and highly significant patterns. It remained, however, for the Maoris of New Zealand to carry body and facial tattooing to the full dignity of a major art.

The habit in India of painting with henna the nails of famous beauties, the toilet rituals of Egypt and old Cathay, perhaps even the ardent love for cosmetics in this day, are all survivals of these same primordial practices.

To transfer this love of color to objects they themselves created, as well as to their own body, was not a difficult transition. The value of a textile fiber would then be determined by the ease with which it accepted color. As soon, therefore, as any advance guard of the human race migrated to a country where cotton grew, the brilliant blossoms and the opening pods of lint would soon attract attention and when it was found that the cotton fiber had a great receptivity for certain dyes, it would become a favorite for its æsthetic rather than its utilitarian qualities.

And so this tiny filament begins its history as one of the many, perhaps at the beginning the least important of the textile fibers. When this was in time, even in relationship to culture, it is impossible to say. It is, of course, possible that these qualities might have been independently discovered by totally distinct races in both Asia and the New World. One thing only is reasonably clear; cotton does not belong in the same class with the first sturdy rough fibers of usage. Its choice was determined because of its qualities as a medium of beauty rather than utility.

CHAPTER III

PRIMITIVE TECHNIQUE

AS cordage and yarn passed from a strictly utilitarian usage into the arts, it became desirable to store surplus product for subsequent use. It was natural that they should wrap this yarn, when spun, about either a stick or a stone, and here begins the two great basic methods of spinning.

Sooner or later the spinner would discover that a stone, wrapped in a covering of yarn, if allowed to hang down and set in a twirling motion, helped the spinning of fiber. This type of spinning was generally practised among wool and flax using peoples and was in existence in Europe and England within historic times, and is even still practised by the Spanish shepherds. (See whorl and distaff spinning illustration.)

It was, however, unsuitable to the short cotton fiber. In this case the yarn was wound on a straight stick. In time the spinner discovered that if one end of the stick rested in a smooth shell or stone, and was twisted with one hand, the other hand might more easily form the thread and smooth out the stick and rough places. (See illustration of Peruvian spinning.)

Practically all subsequent spinning has been developed on this principle. The yarns of Dacca muslins (to be mentioned later), the most exquisite of cotton textures, were spun on this principle, as were those of prehistoric Peru. India later added the spinning wheel to increase the revolutions of the spindle stick and this wheel, some time in the early Middle Ages, was introduced in Europe and became the ancestor of all the spinning devices invented in England during the Eighteenth Century.

A little study of the basketry arts will prove that many of our familiar geometric weaving patterns originated in basketry and were later incorporated in cloth. The same is true of many of the techniques of weaving. There are baskets in tapestry, leno, and twill weaves and in embroidery, all of which subsequently appear in cloth. Consequently, it is little exaggeration to state, that before the first cloth was made, the fundamental methods of ornament, dyeing and fabric construction, had already been developed.

Just as the idea of weaving precedes the idea of cloth making, so does the crudest cloth long precede any type of loom. The loom is, after all, only a convenient implement for weaving, not a necessity, until cloth making reaches its finer phases. Today in the jungles of Borneo, rough cotton hammocks are made without a loom, although the loom is used in the weaving of finer fabrics. The Indians of Northern Canada make a blanket of twisted strips of rabbit fur without knowledge of any implement similar to a loom. The warps in both cases are stretched on the ground and the weft or filling intertwined by hand. With the expansion of ideas in cloth making, some kind of a frame to hold the warp in position during the act of weaving becomes necessary. The purpose of the loom is to keep the warp threads parallel to each other in one plane, at approximately equal distances apart, and to

THE EVOLUTION OF SPINNING

The earliest spinning was merely twisting fibers between the fingers or rolling on the naked thigh. Implements were introduced when the need for finer yarn developed. Implement spinning is divided into two broad classifications—the whorl and distaff method, where a hanging weight attenuates the partially spun thread; and the draught and twist method where the partially spun thread is attenuated through draught created between the spindle and the hand of the spinner. This latter method developed in cotton areas and was finally associated with the wheel to give greater speed to the revolutions of the spindle and hence yield a greater production of yarn.

The first man to imitate mechanically this principle of spinning was Samuel Crompton, who invented the mule in 1779. His home is preserved in Bolton, England, as a memorial museum.

1—Greek woman spinning flax by whorl and distaff or hanging weight method. *(Page 23)*

2—Famous prehistoric Peruvian vase showing spinning by draught and twist method. *(Pages 23, 54)*
American Museum of Natural History.

2a—Aztec mother punishing daughter for poor spinning. *(Pages 23, 40)*
Codex Mendoza
Courtesy of the American Museum of Natural History.

3—Probable method of spinning in prehistoric Peru. *(Page 23)*

4—Hindu woman spinning cotton on primitive wheel. *(Pages 24, 106)*

5—Colonial woman spinning cotton with European model of Oriental wheel with cards and basket of roving. *(Page 24)*

6—Chinaman spinning cotton by adaptation of Indian cotton wheel. *(Page 24)*

7—Hall i' th' Wood, house in which Crompton invented the spinning mule in Bolton, England, 1779. *(Page 114)*

PLATE 3

permit of their easy manipulation during the insertion of weft.

There are two basic types of looms, one which apparently developed in the Mediterranean flax area and the type peculiar to the cotton and silk areas. The simplest loom was the former. This consisted of a single, rigid horizontal bar, from which the warps hung down with weights attached to produce tension and to keep the filatal elements parallel to each other.

This type of loom quickly reached its fullest possible mechanical development. The addition of two parallel slender rods below the loom bar, running under and over opposite groups of warps, further assisted in keeping the warps in their proper relationship and making it easy for the weaver to separate them in weaving units.

Weaving on such a device resembles a kind of embroidery. By no means do I wish to infer that many beautiful webs were not woven on this implement. The early Greeks produced beautiful designs on it; perhaps the earliest Assyrian and Babylonia webs were made on this frame. We know of these designs only from the ceramic pictures of the classical Grecian period and in the degenerate forms of the later Coptic tapestries. This loom did not lend itself to the mechanical subtleties of the more highly involved constructions, and it has no descendants in the machines of today.

Its history is, however, interesting as proving that a mechanical device may have a very wide terrestrial distribution, and be of immense antiquity. We find warp weights in the silt of the Swiss Lakes, judged to be between seven and ten thousand years old. This loom is pictured on the famous Grecian vase illustrating

Penelope weaving the tapestry she unravelled each night in answer to the fruitless prayers of her suitors. In a Scandinavian saga, skulls are referred to as warp weights and this type of loom was used up to modern times in Iceland. So in Europe alone we have a clear record of one weaving device covering areas which were known to use flax and wool of not less than ten thousand years.

To find this loom again we must cross the northern portion of the Asiatic continent over the narrow, island dotted, foggy seas of the Bering Straits, until we come to its last expression among the Haida tribes of Coastal Alaska. Since we know from incontrovertible evidence that the culture of Alaska is of Siberian origin, no one attempts to prove that this particular type of loom was not introduced. It is freely acknowledged as an Asiatic intrusion.

There is no inference of a common blood relationship between the peoples, who over such a long period of time and wide terrestrial area, used these two interesting inventions. I merely cite them to prove how wide a distribution such ideas may have. If such an implement can be traced from Asia to the New World, it at least proves that such things are possible.

No natural fiber is so pliable as cotton and probably among the great primitive yarns none was so weak. Consequently, the uneven tension of warp weights on the single-barred loom, was not suitable. Silk, of course, is almost as pliable as cotton, even in its natural gummed state. It is not surprising, therefore, to find the two-barred loom in early China, as in the early cotton areas of India. Whether one borrowed from the other or both from a common source, there is no possible way of telling. As a matter of fact, India and

China are but modern names given to very ancient areas of human culture.

The earliest form of the cotton loom consisted of two parallel bars, held apart by being attached to other objects. Around these bars the warp was wrapped. Such a loom still exists in primitive parts of Asia and is found in the more inaccessible jungles of South America.

The first improvement was the introduction of two light rods to keep the warps evenly divided and to bring them into a common weaving plane. The next addition was the attaching of the warps to a pliable twisted cord instead of directly to the loom bars, and fastening this cord by a string to the bars. This increased the evenness of the weaving plane and prevented the warps from cutting during the constant movement of weaving.

The last additions to this implement were two sticks with loops of cord attached at equal distances. Through the loops of one, even numbered warps passed; through the loops of the other, the odd numbered. These rods are called healds. The function of these rods is to permit the weaver to divide alternate warps in equal groups, for the convenient insertion of weft in weaving. The weaver lifts one rod with the attached warps and thus forms a triangular space, through which to insert the weft. Releasing this bar, he lifts the second one and forms another triangular space or shed with the second group of warps. By alternating this movement and inserting the weft in each shed, a woven fabric can be made.

There are certain minor additions of tools, such as a heavy stick of polished wood called a weaver's sword or battern for beating each pick of weft closely to its predecessor and lighter daggers of polished wood or bone to insert at right angles, between two warps to produce more compact fabrics. In time, a comb-like implement, which operated on a larger number of warps, was invented to take the place of the weaving battern and dagger.

In India, this form of loom had a further addition. The heald bars were formed into a frame known as heddles, attached by a cord running over an overhanging bough of a tree and by loops to the feet of the weaver stretched beneath the warps. (See picture of Indian loom.) The separation of the warps into weaving sheds was performed with the movements of the feet. This loom is simply a mechanical improvement over the earlier type I have described. Undoubtedly, this form of loom was introduced into Europe from Asia Minor at a very early date. The Fourteenth Century English silk loom, illustrated in the text, is evidently the same in character as the Indian loom. If silk and cotton were introduced at about this time, or a little earlier, and the spinning wheel, it is not difficult to understand how the loom was borrowed also.

The basis of all modern looms is the two-barred principle of warp arrangement and shedding with heddles and harnesses.

I hope in this technical discussion of the development of the primitive loom I have disabused the reader's mind of the idea that it is simple in the sense of lacking in breadth of intellectual conception. It is in fact one of man's greatest technical achievements. On this type of loom every cloth we know today, every weave in the history of cloth making has been produced. Peru, perhaps the greatest of textile people, of whom we have an accurate material record, never developed the loom beyond the hand type.

PRIMITIVE TECHNIQUE

THE DISTRIBUTION OF THE WARP WEIGHTED LOOM

The warp weighted loom was first developed in the flax area of Europe. The oldest type was discovered in the Neolythic Swiss Lake villages, and is judged to be about ten thousand years old. This loom appears in classical Greece, Scandinavia, Iceland, and was perhaps the loom used in Britain before Cæsar's time. Its latest form is found among the Haida Indians of the Alaskan Coast and is an intrusion from Asiatic migration. *(Pages 25, 35)*

WARP WEIGHTED LOOM

1—Warp weighted loom from Swiss Lakes. *(Page 26)*
2—Warp weighted loom from Scandinavia. *(Page 26)*
3—Greek loom with Penelope and the suitor from Greek vase. *(Page 26)*
4—Warp weighted loom from Iceland. *(Page 26)*
5—Haida warp weighted loom from Alaska with ceremonial apron in tapestry weave. *(Page 26)*

THE TWO BARRED LOOM

The distribution of the two barred loom is discussed in the chapter on the New World. In both Asia and the New World, it is apparently associated with the history of cotton. This loom is not indigenous in Europe, but was introduced from Asia at a very early period. *(Pages 25, 35)*

1—Diagram of a Peruvian Tapestry Loom. a, a', Loom bars; b, Weave dagger forming short shed; b', Weave dagger beating up pick of weft just delivered by bobbin (d); c, Bobbin of weft being drawn through shed formed by (b); d, d', d'', d''', d'''', Bobbins containing the different colors of yarn required in fabrics; e^1, Warp twisted from small groups to avoid tangles; f, Yarn from bobbin (d') closing up slit in weaving; e, Shed formed by weave dagger (b). *(Page 33)*

2—Drawing of tapestry weaving with design notations. Introduction à la *Histoire Antigua del Peru*. Dr. Julio C. Tello, Lima, Peru. *(Page 33)*

3—The Common Type of Peruvian Loom. a, a', Loom bars; b, b', Loom strings; c, c', Binding strings; d, Weave sword beating up weft; e, Warps not attached to heald rod (f), hence not lifted; f, Heald rod lifted to form shed; g, Warps attached to heald rod (f) and raised to form shed; h, Weft just delivered by spindle; i, Spindle after inserting pick of weft (k); j, Fell of cloth (already woven portion of web). *(Pages 27, 33)*

4—Prehistoric Peruvian Loom with partially woven double cloth. *(Page 27)*

NOTE: Compare this with technical drawing number two. American Museum of Natural History.

5—Hindu weaving on loom with heddle frame to separate the warps. *(Pages 26, 28, 66)*

6—Two barred loom with heddle used by silk weavers in England in the Fourteenth Century. *(Pages 28, 107)*

7—Chinese loom for cotton weaving. *(Page 26)*

PLATE 4

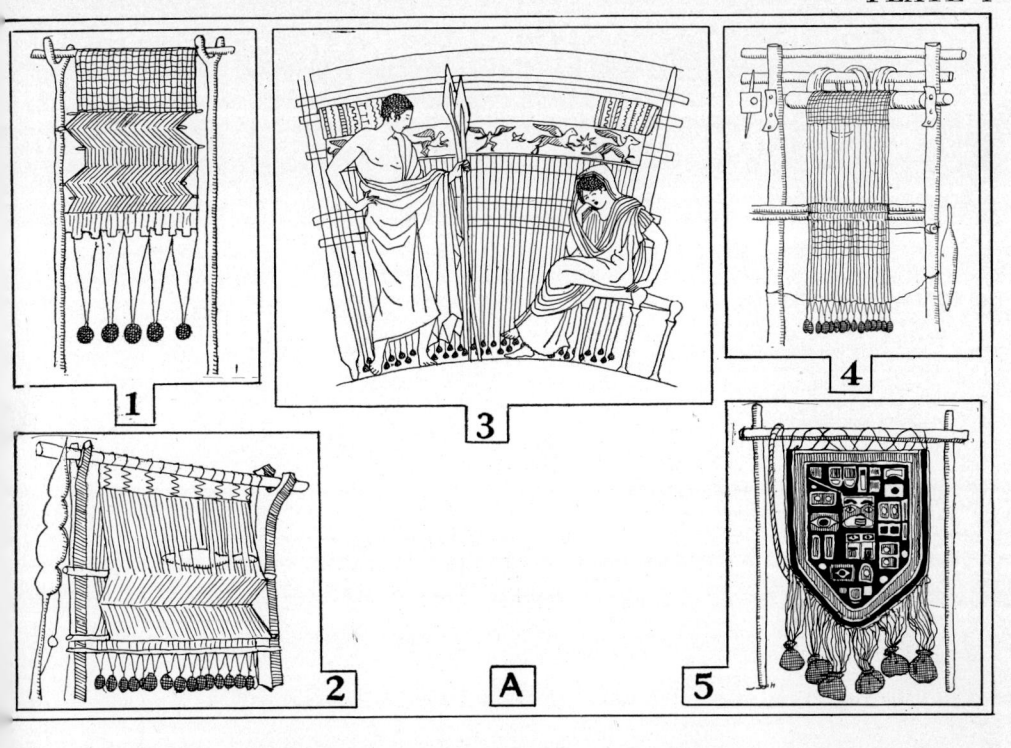

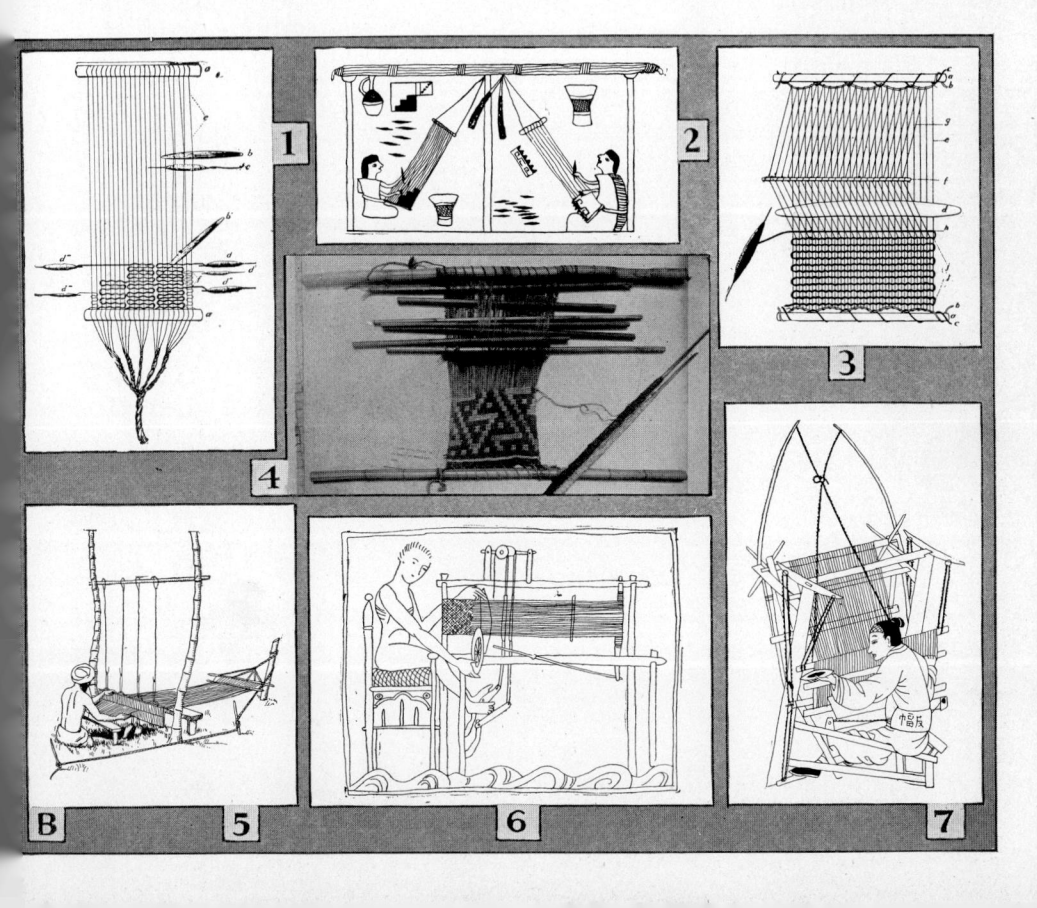

Both of these types of looms, the true hand loom and the foot treadle loom, are used in Europe to this day. The hand loom appears among the rug weavers of Asia Minor and the tapestry weavers of Europe and the foot treadle loom without the addition of power, is used for the production of the finer fashion fabrics in France and Switzerland and is even finding a footing in America.

The fact that this type of loom originated in India and spread to Europe, that we find it together with the technical subtleties of fabric construction in the cotton area in the New World, is difficult to explain except on the assumption of direct or indirect social contact.

CHAPTER IV

THE NEW WORLD

THE modern story of cotton in the New World begins with the landing of Columbus in the Bahama Islands in the Fall of 1492. To his delight he beheld the natives wearing cotton garments. This could only mean that he had reached the Indies, since to his mind any land producing cotton must be the golden Orient.

At the time of the Discovery, only two European peoples were familiar with the East and its products. The Spaniards knew of the Orient through six hundred years of constant warfare with the Moors; the Italians through fruitful centuries of trade and intercourse. Columbus was an Italian, not unfamiliar with the ports of Asia Minor. He sailed on this memorable voyage as a Spanish captain of fortune. Hence the finding of cotton had a double significance for him.

It was the twelfth of October when he landed, the harvest season for cotton in this latitude and fairer than any earthly flowers to our great Genoese must have seemed the white streamers of the opening bolls. The first American cotton, therefore, to reach the Old World from America, was brought back by the successful dreamer as proof to the skeptical court of Ferdinand and Isabella.

It is one of the curious accidents of history, that practically all subsequent Spanish discovery and conquest in the New World was among peoples well advanced in the arts of cotton spinning and weaving. Cortez, Pizarro, Balboa, Alvarado and Coronado, each came in contact with peoples distinct in culture, if basically similar in race, yet all were skillful weavers of cotton. With the unimportant exceptions of Florida and the moist delta of the Mississippi, all of Spain's vast Colonial empire of the Sixteenth Century in the New World was carved from the prehistoric cotton area.

The prehistoric cotton map of the New World extended southwest from the middle of the State of Utah through the desert parts of our Southwest, Mexico, Central and South America. It did not include any of our Central Mississippi or Atlantic Seaboard cotton states of today, although it may have included the southwestern fringe of Texas.

Within this ancient region we have incontrovertible evidence that cotton culture existed for many centuries. In most instances it perhaps antedates the Christian Era and in all locations there is reason to believe that our previous notions of antiquity fall far short of the actual truth. Not all of these people, however, achieved an equal skill. The art seemed to have moved northward, our Southwest being its uttermost fringe. It either originated in prehistoric Peru and spread northward, or in some region in Central America and spread in both directions north and south.

In these regions we know that great civilizations reached maturity and passed into decay long before the Spanish conquest and in both regions the cultivation and conversion of cotton were among the earliest and most advanced of the arts. Among the Mayas of Central America a date of 632 B.C. has recently been firmly established through the calendic researches of Dr. Herbert J. Spinden. But people who had arrived at a sufficient culture to have developed an accurate calendic system based upon the observation of astronomical bodies, must have reached a very high degree of culture, suggesting an infinitely remote past. This civilization developed in a climate that did not permit the preservation of fabrics, except in a few special instances.

The fact that the perfect desert of coastal Peru has preserved for us each priceless web, each subtlety in design and technique, each implement and tool and process, and the moist climate of Central America has destroyed almost every direct evidence of the fabric arts makes no difference in the rival claims for antiquity of either region. These are but historic incidents, and there is no direct proof as yet to establish one or the other as the more ancient.

Be its point of origin, however, where it may, all over this vast and contiguous area, peopled by tribes of common racial stocks, there is unquestionably a similarity in technique, design and implements, and a universal use of cotton. It is one story retold in many forms and across many centuries.

The distinctions in the ancient arts are themselves an eloquent proof of antiquity, since in all these areas the tools, implements, fibers and in many instances the dyes are identical. No one familiar with the different areas would have the least difficulty in distinguishing the art of one region from that of another. There is, however, a similarity as well as a difference. This can be accounted for by the fact that in realistic designs primitive people represent animal forms, more or less modified to suit their limitations of expression and the advancement of their spiritual conceptions. Since the fauna through all this region is practically identical, with the exception of the high Andes, we must expect occasionally parallels in draftsmanship. There was as well an intermittent trade between certain regions proven by the fact that the arts of one region are sometimes found in the ruins of another.

The closest similarity in design is, however, in geometric patterns. All of these people were skilled textile workers, and the geometric limitation of the loom was familiar to each. Wherever in this vast area, covering portions of two great continents, cotton appears, there also is the two-barred loom described in the preceding chapter. The loom of prehistoric Peru, from eighteen hundred to three thousand years old, is exactly the same as the loom discovered within the last fifty years among the Huichol Indians of Southern Mexico, and everywhere else, where we find cotton until the Spanish loom was introduced. No change whatever has taken place in the loom in all these centuries; it has remained absolutely static. The correspondence between the two-barred loom and cotton plant is absolute. Wherever this type of loom appears, there is cotton, and wherever cotton appears in ancient times, this type of loom occurs.

Certain it is that the two-barred cotton loom and the fundamental fabric constructions, which seem inseparable from this implement, must have spread from one single source in America. So complicated and perfect an instrument, so profound a technical knowledge can not have had in one connected area dual or independent origins.

No very definite theories have yet been advanced by anthropologists to account for the great archaic cotton cultures in the New World. Their antiquity is conceded by all, of course, and probably exceeds any of the cautious estimates.

In a general way, Asia is recognized as the remote cultural home of the Americas. Since there is incontrovertible evidence of migration from Siberia, even up to comparatively modern times, into Alaska, there is a rather hazy belief that this single point of contact explains all human life on the two continents of North and South America.

I have no intention of discussing this problem any further than to call attention to certain facts directly pertinent to this narrative, which seem at variance with this theory.

The famous sinew back bow, which originated somewhere in archaic Mediterranean cultures, probably Assyria, has a distribution which corresponds rather closely with that of the warp-weighted loom. Neither the warp-weighted loom nor the sinew back bow occur in any of the cotton areas of the New World. At the northern-most fringe of cotton, the bow was obviously borrowed from the non-cotton, nomadic tribes.

Among the Aztecs, the bow and arrow was the hieroglyphical symbol of the wild tribes. Their own weapon was the throwing stick, a prototype of the bow. In the elaborate and beautiful stone carving of the Mayas of Central America there are no representations of the bow. The ancient Peruvians, in whose sandy graves we find so perfect a record of their arts, were unfamiliar with the bow. The sling was their missile weapon. The bow does occur as a plain stick among the tribes to the South and generally through the jungles east of the Andes. There is, however, a rather strong belief that these peoples were not alone far inferior in culture, but actually distinct in race from the higher races of the cotton areas.

South of the culture of Alaska, where the warp-weighted loom prevailed, stretched the great American plains where for a long distance no loom of any type is found, but where the bow does exist and has been known for a great period of time. The northern peoples still adhere to the powerful, composite sinew back type which gradually fades out to the single stick type.

It is, of course, possible that a slowly drifting migration of peoples, coming in contact with highly varied terrestrial environment, might easily have in the cycles of time lost the warp-weighted loom. The weaving of the cloth suggests a static culture, and cloth itself is a luxury not a social necessity. I am, however, extremely reluctant to believe that any peoples living largely by the chase would have discarded so perfect a weapon as the bow.

If, therefore, the peoples of the pre-Columbian cotton areas came at some very remote time from the North, we must determine how they came to change their type of loom, after losing it entirely, and how they abandoned the bow once they reached the area of cotton.

Were there then two great roadways from Asia to the New World? Were the ancient peoples of Mexico, Central America and the Pacific Coast of Peru of distinct racial stock from the red men of our plains, forests and frozen tundras? This is by no means proven by the few facts I have outlined. The point is, however, clearly raised, and if it may not be affirmed, neither can it be denied. Nor is there lacking supporting evidence in the general structure of common myths, in the presence of similar objects in both continents, as well as in the curious relationship of loom types, weave techniques and the cotton plant itself.

A glance at the world map will show a most significant distribution of islands in the Pacific along the Tropic of Capricorn. It is well known that these islands are in most instances really the peaks of submerged mountain chains. In ancient misty ages this chain may have been more closely knit. Among these island peoples there are legends of long journeys to unknown continental coasts, that are suggestive of the legends of the Atlantic Norsemen before Columbus. The survival of certain arts which are strongly reminiscent of textiles, although these peoples are unskilled in loom-work, may be additional proof. Their tapas or pounded bark mats are generally geometric in design and this characteristic in ornament we naturally and rightly associate with some form of weaving.

In the next chapter I will present proof that clearly establishes southern India as the home of cotton so far as Asia, Africa, the Islands of the Indian Ocean and Europe are concerned. Unless we are to assume this miracle of ingenuity had two distinct manifestations, we are compelled to acknowledge that the cotton technique of India and the New World must have been derived from some common and as yet unknown source.

The most temperate presentation of such evidence is capable of misinterpretation highly prejudicial to the clear reasoning of science. So great a problem as racial and cultural origins can not be settled on a basis of any single group of facts, however broad these may be. Enough has been suggested, however, to make it evident that the broader problem and the complete history of the cotton plant and cotton techniques have much in common.

For many years careful investigations have been conducted in our Southwest. The ruins of many ancient people, who lived and prospered centuries before the Spanish conquest, have been scientifically excavated. Among all these people cotton was used and we find remnants of cloth, seed, lint, even spindles and ancient looms. Cotton among these people, however, had a religious rather than a utilitarian usage. The basic fiber of this area was the tough yucca grass. Cotton was an intrusion from a greater civilization to the southward, and is used to this day in most of the religious ceremonies retained by the descendants of these people.

Many Hopi ceremonies are associated with the prayers for rain, and symbolic altars are raised to the mystery of fertility. The rain clouds are represented by billowy masses of white cotton and the miracle of the falling rain symbolized by the straight cords of the unwoven upright warps stretched on the loom. The dress of the Hopi bride is cotton, grown, gathered, spun and woven by the relatives of the groom. These are but two instances of the importance of cotton in the religious ceremonies of these peoples. It is doubtful, however, if it had any broad secular use, until modern times.

The few fabrics of cotton discovered in the dry caves and burial sites of the Southwest, give us a rather high estimate of their skill. One tapestry apron, a part of which is in the American Museum of Natural History and another in the Museum of the American Indian, is a splendid example of craftsmanship. This apron was found wrapped around a naturally desiccated human body, discovered in a dry cave in Grand Gulch,

Utah, and is believed to antedate the Christian Era. In the Brooklyn Museum, there is a child's knitted shirt of unique pattern and a few fragments of stencilled cotton duck, which are of equal importance in establishing the technical skill of these people. There are as well woven patterns not unlike our own modern brocades and embroidery among these relics.

From the evidence furnished by material culture cotton had a peculiar and important place in the life of the American Indian in the southwestern United States. Its antiquity is assured by its frequent presence among the remains of the ancient Cliff Dwellers. There are hanks of white cotton cord and of dyed cotton cord in the collections from the cliff dwellings of the Cañon de Chelly, Arizona, in the Brooklyn Museum. This cord, from the objects found with it, appears to have formed part of the contents of an aboriginal work basket. In the same collection is a child's garment made of cotton found on the body of a child in the White House ruin in the Cañon de Chelly. With this garment are fragments of a dyed cotton blanket showing a pattern and cotton tassels with which these objects were ornamented. Yucca fiber, prepared from the fresh green stalks by chewing, was the ordinary textile material of the Cliff Dwellers and this garment and blanket were no doubt exceptional and the work of some devoted mother. Among the existing Zuñi and Hopi Indians, in common with the other Pueblo tribes, cotton was used for ceremonial purposes, pointing to a tradition of high antiquity. The ceremonial white garments worn by the manas or unmarried girls were made of cotton as were those of their representatives in the dances. The Hopi girl's wedding dress was of white cotton. The cotton blankets had cords at their corners with which they were tied. These cords were wrought into the forms of ears of corn on the embroidered cotton blankets worn in ceremonies. The woven cotton belt with long fringe at the ends which was used constantly in Pueblo ceremonial costume was also made of cotton. These cotton blankets and belts were special objects of barter among the Pueblo Indians. They corresponded with money, having not only a fixed but a high standard of value. Cotton only was used for a great variety of ceremonial purposes by these Indians, no other fiber being regarded as having any efficiency. It was, indeed, a magical substance. For example, the prayer sticks were tied in pairs with it and it was used to attach to them the ceremonial "breath" feather which was associated with their potency. The tipony or magical wand of feathers was tied with cotton cord as were indeed all the ceremonial sticks used in medicinal ceremonies not only by the Pueblos but by the Navajo as well. Consulting authorities, it is believed from lack of mention by early writers that cotton was not cultivated by the tribes of the southern section of the United States and that the cotton blankets seen by De Soto's troops on the lower Mississippi were brought from the west, possibly from the Pueblos. The Hopi are now the only cultivators of cotton and the robes, kilts and scarfs which they make find their way by trade to other tribes who employ them in their religious performances. In the time of Coronado (1540–42) and Espejo (1583) cotton was raised also by the Acoma and Rio Grande villages in New Mexico. The Pimas in southern Arizona also raised the plant until about 1850 when the industry was brought to an end, by the traders who introduced cheap fabrics, except among the Hopi. In ancient Zuñi and Hopi mortuary rites raw cotton was placed over the face of the dead and cotton seed deposited with the food vessels on the graves.

This is indeed a scanty record, compared with that of more favored regions, still sufficient when taken in connection with their pottery, to prove that their artistry and craftsmanship were of respectable order. Opinions differ and naturally change with the development of new evidence, but the general consensus of opinion is that cotton was introduced in this region prior to the Fifth Century B.C.

In spite of our great knowledge of the Aztec culture and the dramatic interest aroused by these vigorous people in their contact with the Spaniards, we have, comparatively speaking, few examples of cloth that we may safely date from the pre-Spanish period. The climate in Mexico is generally inimical to the preservation of fabric, nor have we as yet scientifically and systematically investigated all possible grave sites in this turbulent sister republic. But the accounts of the Spanish conquerors and the administrators who followed them, beginning with Cortez in 1519, the designs on their pottery, wood and stone carving and the picture writing in their marvelous codices give us incontestable evidence that they were well advanced in the arts of cotton and that cotton was the fiber chiefly used by them for textile purposes.

The famous tribute roll of Montezuma gives us an accurate record of the annual assessments levied in bales of cotton and bolts of cloth and decorated blankets by the warlike Aztec Confederacy upon the tribes who acknowledged the ancient city of Mexico as overlord. One page shows forty bags of cochineal and two thousand decorated cotton blankets, together with many other objects of value as the annual assessment of eleven tributary cities. It has been estimated on a basis of modern values, that this entire tribute ran into millions of dollars each year.

The Spaniards were quick to recognize the skill of the natives as weavers and organized them into factory groups and annually exported to Spain immense quantities of fabrics as well as cotton fiber. The practice of forcing an out-door people to work indoors, together with the brutal slave-driving methods of the Spanish task masters, was so detrimental to the health of this people, that in the famous humane Laws of the Indies, promulgated by the Spaniards in 1540, the practice was forbidden.

One of the most important archæological discoveries in the New World was recently made in the little crater lake at Chichen Itza. This pool of deep, clear water, surrounded by high cliffs of limestone, was, according to an ancient tradition, once the scene of dramatic human sacrifices. These sacrifices had been discontinued long before the Spanish invasion, but none the less lived in the native legends. A recent scientific expedition dredged this little pool with a modern steam shovel and the discoveries were amazing to a degree. In the silt and mud were found beautiful repousser gold breast ornaments, bits of carven jade ornaments, (broken that their spirits might go to the gods), incense bowls filled with copal or rubber gum to produce the white and black smoke so beloved by the Upper Powers. These were studded with jade beads, in some instances only a single jade bead surrounded by green beans, for the priest had discovered that for all the malignant powers of their gods, they could none the less be deceived with impunity.

The least imposing of these discoveries, but by far the most important, as far as this narrative is concerned, are little fragments of charred cotton cloths, once parts of the garments of the deluded and unfortunate victims. These fragments are in far too fragile a condition and far too precious to permit of exhaustive analysis. Fortunately, this is not necessary to determine their technical character, since the weaves are open and easily classified without dissection.

All color has naturally disappeared due to the action of the water and the acids in the silt during the centuries of their immersion, but the types of fabric remain clear and unmistakable. They are the same as those found in other of the cotton regions, particularly in Peru and even among the living primitive tribes of the same general region. If a full collection might possibly be gathered, they would no doubt include all the techniques. So far I have examined brocades, crepes, ducks, and embroideries from this source. All of them give an evidence of a high technical skill, although not particularly fine in weave.

Among the modern natives, living in isolated regions and comparatively free from Spanish and later European influences, there is still a high skill in fabric construction and cotton remains the principal fiber. Here and there in these out of the way regions, ancestral crafts are practised in something approaching their pristine splendor, but the mixture of trade yarns and trade cloth and the incorporation of Spanish designs make this record of doubtful validity. The intrusion of Spanish design, and in some measure of Spanish technique, begins very early in the Sixteenth Century, almost at the very dawn of the conquest, and was so penetrating that it has often confused scholars, studying the native art, who were unfamiliar with Moorish and Spanish motifs. Indeed the natives themselves have no very clear idea of the comparatively recent origin of their present arts. Three centuries of usage among an illiterate people is very likely to obscure sources of origin.

It is but fair to state that these intrusions were not always intended to harm the natives, nor were they actually harmful, except from the scientific point of view. Most of the native arts were associated with their religious ceremonies and against these, the Jesuits quite naturally and sincerely waged a constant warfare. But back of this was the wish to fit the conquered peoples into the life and arts of Europe and make them an integral part of the Spanish Empire. There can be no question of the sincerity of the Jesuit missions in this purpose, however mistaken it may have been. And it must be admitted with frankness that no people in Europe with the possible exception of the Italians were at this period so rich in decorative arts as the Spanish. They had not alone a reminiscence of their own Gothic period, but a great enrichment from their recent Moorish enemies and it rises as a white column to the memory of these devoted men that there still survives through Central, and South America some of the most efficient craft schools in embroidery and weaving in the world today.

One example of how powerful was the Spanish intrusion may be found in the Huichol fabrics from the highlands of southern Mexico and the closely related Cora tribes. These people were conquered according to Spanish history in 1700. Fifty years later they revolted and drove out all Spanish priests, soldiers and administrators alike, but retained Spanish designs in their characteristic double cloth weaving. Here occur the vigorous drawing of Christian saints, Moorish conventions and bird forms originally created by the master weavers of the Byzantine looms, adopted by the Saracens, and so by the Spaniards brought to this remote and alien people.

The San Blas Indians, on the coast of Central America, have stoutly maintained their isolation from white contact. They are a war-like, self-contained people living in an inaccessible coastal region, where fortunately for their security neither oil nor gold have yet been found in any quantity. They import certain trade cloths in small quantities, but their chief demand on civilization is for high powered modern rifles. Their applique work, using two colors of fabrics, is very interesting. Their patterns, strong and vigorous in composition, suggest, however, a degeneration from some higher and forgotten technique. One little jacket of gray cotton cloth with a design worked out in aigrette down, twisted into the weft, has a peculiar historic interest. The first landing of Columbus on the mainland was near this region. He describes the natives as wearing cotton garments with designs formed from the inter-weaving of feathers. For a long time scientists assumed that this was a careless description of the feather ponchos, similar to those of Mexico and Peru. As a matter of fact this little garment proves that the great navigator accurately described the costumes as he saw them.

We know very little of native dyes. Cochineal was cultivated in both Peru and Mexico from immemorial time and this yielded so beautiful a red, that the Spaniards introduced the cultivation in southern Spain at a very early date, and it spread into the Near East,

NEW WORLD: SPANISH INFLUENCE
PART 1

Almost with the Spanish conquest the primitive arts of the New World were affected. The Spanish priests desired to destroy all records of the ancient pagan religions and the Spanish governors took advantage of the native skill to make merchandise for the Spanish market and to supply the demands of the Spanish residents in the New World. The intrusion resulted in many very beautiful forms and the arts of Latin America today are largely the result of these contacts. *(Page 43)*

1—Loom from Cora Tribe with partly finished double cloth showing Spanish design with native drawing of horses. *(Pages 33, 44)*
Museum of the American Indian.

2—Huichol double cloth bags with designs of the double eagle of the Holy Roman Empire. *(Page 43)*
American Museum of Natural History.

3—Detail of Mexican embroidery with Saracenic motives. *(Page 43)*
Museum of the American Indian.

4—Detail of Hispano-Inca tapestry shawls with mixture of Spanish and Inca designs about 1570. *(Page 43)*
Museum of the American Indian.

5—Detail of Inca poncho with border of Spanish figures worked in tinsel yarns about 1550. *(Page 43)*
American Museum of Natural History.

6—Corner of embroidered Mexican scarf with Moorish motives. *(Page 43)*
American Museum of Natural History.

7—Detail of Hispano-Inca shawl with Spanish motives worked by Inca craftsmen about 1580. *(Page 43)*
American Museum of Natural History.

PLATE 5

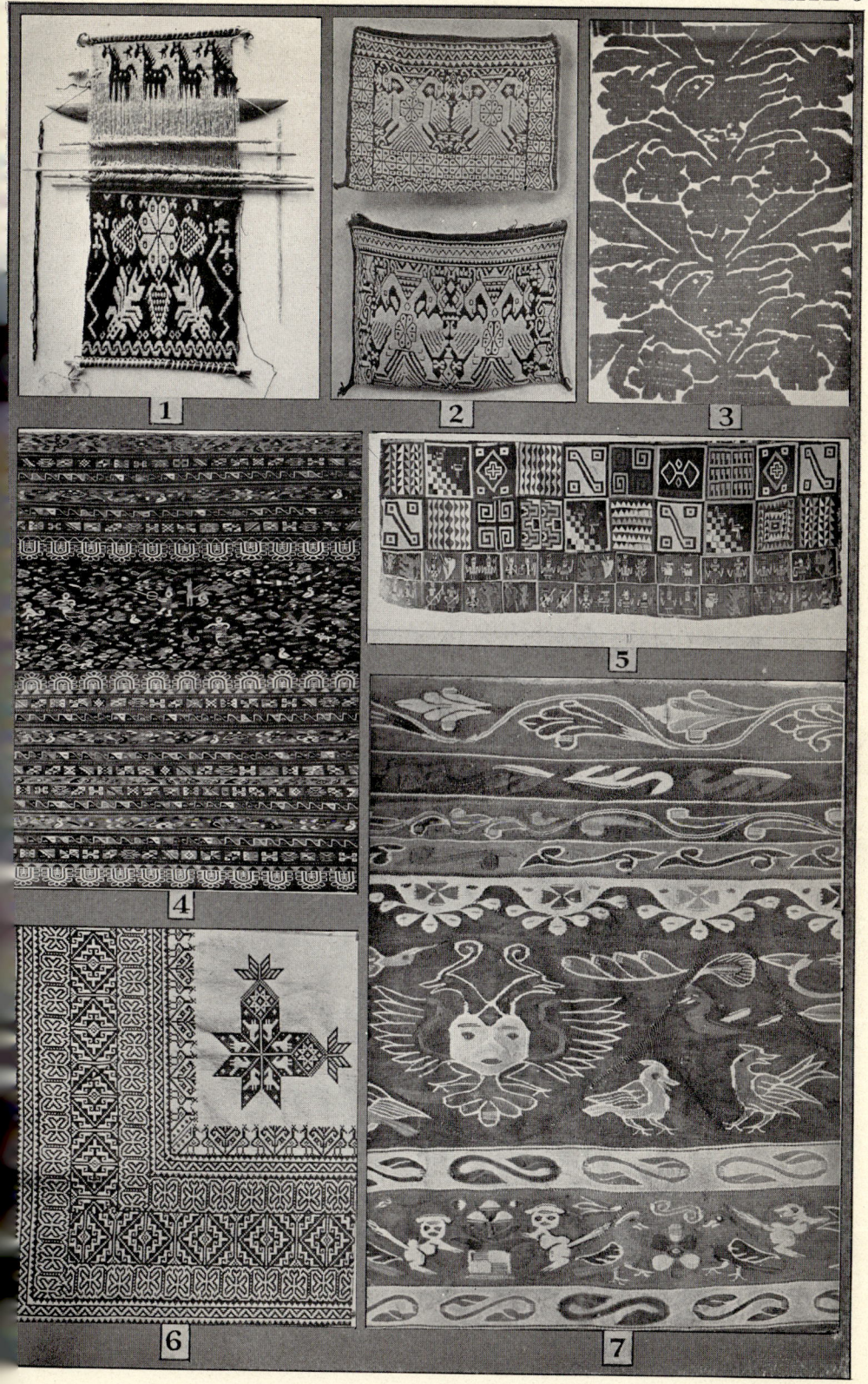

where it has remained ever since. Of peculiar interest, however, is the dye from the purpura shell fish which produced the beautiful purples still found in certain of the native cotton fabrics and which was unquestionably the basis of the fine purple shades in the ancient fabrics of Peru. This little mollusk is found on the coast of Central America and in a small patch off the coast of Peru. The natives wade out in the shallow waters and squeeze the juice into a shallow dish, and then return the mollusk safely to its native element. The liquor is colorless, although not entirely free from odor.

All over the regions that I have briefly covered, the cultivation of cotton and the arts of weaving and dyeing cotton fabrics existed for many centuries before the Spanish invasion and in most instances before the Christian Era. The same type of looms, the same general techniques pervade all of these regions, and form obviously the same story, told in many different ways. Only recently have our great museums begun to carefully collect the materials from modern tribes and clarify their existing traditions in comparison with our rapidly growing knowledge of the archaic cultures, but as far back as we can trace, the story of either the arts or agriculture of these people, we find it indissolubly associated with the still more ancient cotton cultures.

CHAPTER V

PERU

NOT the least interesting phase of the history of the fabric arts of pre-Inca Peru, is the fact that we may study the technical and artistic development of a single people without the confusion of outside influences or intrusions. No matter what may have been the remote origin of this mysterious people, it is beyond question that their civilization is their own, developed through many centuries in a single environment from the spiritual and material reactions of a people of a common or homogeneous race.

It is the most perfect fabric record left by any people in the history of the world. Here we do not have to rely on conjecture, traditions or the comparison of related arts, but may study in all their variety the actual fabrics, tools and implements, and through these safely reconstruct the processes and methods. No people in the history of fabrics, in any part of the world, have ever achieved such a high technical skill nor excelled them in conception of design, composition or the use of color.

Beyond question, the most ancient specimens of cotton fabrics in all the world are those found in the desert graves of pre-Inca Peru. They may even be older than the earliest mention of cotton in the records

PLATE 6

NEW WORLD
PART 2

In very few places outside of coastal Peru did climatic conditions permit the preservation of cotton fabrics. Our knowledge of their undoubted skill must, therefore, be taken from related arts, Spanish accounts at the time of the Conquests and the few actual specimens preserved through fortunate accident.

1—Aztec Chief wearing cotton robe. Codex Rios. *(Page 40)*
Courtesy of American Museum of Natural History.

2—Aztec noble wearing cotton shawl. Codex Don Fernando de Alba.
Courtesy of American Museum of Natural History. *(Page 40)*

3—Aztec noble wearing cotton garment. Mexican Codex. *(Page 40)*
Courtesy of American Museum of Natural History.

4—Aztec women wearing cotton garments. Mexican Codex. *(Page 40)*
Courtesy of American Museum of Natural History.

5—Oldest cotton tapestry found within the limits of the United States. From Grand Gulch, Utah. *(Page 37)*
American Museum of Natural History.

6—Aztec two barred loom from Mexico. Mexican Codex. *(Page 40)*
Courtesy of American Museum of Natural History.

7—Cotton blankets or shawls paid as tribute, ancient Mexico. Mexican Codex. *(Page 40)*
Courtesy of American Museum of Natural History.

8—Terra cotta stamp and cylinders for printing paper and cloth, ancient Mexico. *(Page 40)*
American Museum of Natural History.

9—Detail of Hopi bride's cotton robe. *(Page 39)*
American Museum of Natural History.

10—Modern cotton jacket with interwoven feather pattern of the type mentioned by Columbus, San Blas Indians. *(Page 44)*
American Museum of Natural History.

11—Prehistoric incense bowl, Central America, with design executed in resist dyeing. *(Page 59)*
American Museum of Natural History.

12—Prehistoric stencilled cotton, and knitted cotton garment from Cliff Dwelling Cañon de Chelly, Arizona. *(Page 38)*
Brooklyn Museum.

13—Maya women wearing designed cotton costume, Yaxchilan, "Maya Art." *(Page 32)*
Dr. Herbert J. Spinden.

PLATE 6

of India, which occurs in the first millennium before the Christian Era, although not older than the first actual appearance of cotton fabric in the East, which naturally precedes its first literary mention by many centuries.

It is impossible, however, to give any reliable dates for this great civilization. We can only gain some comprehension of its antiquity by a comparison with our own historic experiences.

Roughly speaking, the people conquered by Pizzaro in 1532, and known as Incas, were at about the same cultural level as our own European ancestors of the Twelfth Century. Our ancestors indeed excelled in the knowledge of iron and steel, but the Incas were infinitely more advanced in road and aqueduct building and in agriculture. Our ancestors had a written language, but the Incas had a political and social system that had abolished those terrible famines and plagues, which at times wrought such havoc in old Europe. In architecture a reasonable judge will yield the palm to South America, since our great Gothic period comes after the age I have mentioned. And so far as comforts and luxuries of life are concerned, there can be no comparison.

In this connection it is but fair to admit, that our forefathers built their culture on the vigorous remains of Roman civilization. After the Crusades, they were open as well to the influences of the later Greek civilizations and powerful suggestions from Oriental sources. Most of our arts, many of our forms of law, the dawn of modern science and language show traces of highly diverse and distinguished sources of origin. It is no accident that Latin, Greek, Arabic and Hebrew are still regarded as the classical languages which lie back of our civilization. In other words, ours is a highly composite civilization, with all the immense advantages of these inspirational and guiding factors. Consequently, he who would seek surely for our beginnings, can not be guided wholly by arbitrary, national dates. Our civilization to a degree is only our own through adoption and intrusion, not through creation.

Our textile arts, especially as these are concerned with silk and cotton, were borrowed almost in toto from the East. Design, technique, even philology, indicate how vast is our debt to these alien peoples. Yet in spite of these vast advantages, in spite of the slow accretion of the ripening centuries of our culture, in some ways even today we have still to reach the achievements of this mysterious people of the Pacific Slope of the Andes. Consequently in estimating their age, we must either attribute to them miraculous mental powers in advance of our own, or acknowledge that their civilization must have been the ripened fruit of longer periods of time, than any scientist has yet dared to estimate.

It must be remembered as well, that the Incas themselves, for all of their venerable age, were newcomers in this region. Their traditions, gathered by the Spanish sons of Inca mothers and written within a half century of the Conquest, record their first appearance on the coast from their mountain homes of the interior. There is little doubt, that they were of the same race as this still older people, but had been separated from them for many centuries and had lost all contacts with them before their exodus to the coastal regions which occurred, according to these traditions, about two or three centuries before Pizzaro and his armored men landed on their shores. Consequently in dating these cotton fabrics, we have to take into consideration, not only the age of the Inca civilization, but of this still more ancient culture.

It is generally acknowledged that all the peoples of the Western Hemisphere, must have come originally from Asia, at a very remote time. The most conservative estimate of the age of this intrusion, is twenty thousand years or before the last Ice Age. Botanists and biologists do not agree in this opinion, but push it backward into a still more remote antiquity. These figures are at best but guesses and concern themselves with the length of time it must have taken to produce certain food plants and domesticate certain animals. We know at least, that the so-called Dawn Man has never been found in the New World. There are no skeletal remains such as the famous Piltdown skull or the Java-man to push our human record backward into pre-culture ages. Consequently when man migrated from Asia to the New World, he brought with him some rudiments of culture, since such a journey presupposes at least a crude measure of social organization, but this culture must have been of a very low order indeed, some tools and knowledge of processes, perhaps a hazy understanding of agriculture, nothing more. To all intents and purposes, the ancient civilizations of the New World are indigenous, the fruit of experiences and achievements in this new home.

How long, therefore, must it have taken these people to reach such a degree of culture, to have grown to maturity and to have vanished completely? These considerations are our only guide in establishing comparative dates, which are after all of little significance in such a history, our penchant for them, resting largely on the predominant part that genealogies play in the chronicles of Europe. We must measure history in Peru by cultural achievements, a far surer gauge of value.

When the Incas first descended from their mountain home, they came upon the vast, deserted ruins of a vanished people. In the diamond dry air of this region, material decay is apparently inoperative. It is another and more perfect Egypt, an Egypt never in danger of an overflowing Nile. Yet the Inca van guards, drifting to the coast, found ancient buildings, among the world's most amazing architectural records in ruins, massive blocks of stone eroded even in this preserving climate.

These were an agricultural people, supporting according to estimates, in ancient times, twice the modern population of the countries once included in this empire, Peru, parts of Colombia, and Bolivia and the northern sections of Chile. Agriculture was only possible through irrigation and the only water came from the melting snows and ice of the impassable Andes. Consequently every bit of ground where cultivation could be conducted, with the most Herculean efforts was utilized. There are records of the natives clearing away twenty feet of sand to reach alluvial soil. Their grave sites were located therefore, in particularly dry sections of the desert. Thus delicacy of technique and brilliance of color have been preserved for us since it was their custom to bury with the dead all personal belongings, tools, implements and unfinished work.

It is but justice to the Spanish memory to write that the great cultures of America, in the Mayan area and in Peru had already passed into decay centuries before the Spanish invasion. Against their record can not be charged the destruction of the greatest civilizations of the New World. The peoples, whom

PERU
PART 1

1—Preparation of cotton for spinning. *(Page 54)*
    A & B—Cotton with seed removed.
    C—Cotton carded and felted into lap.
    D—Cone of carded fiber.
    E—Tips of cone and spindle with roving attached at the beginning of spinning.
    F—Famous vase showing spinning.
American Museum of Natural History.

2—Spinning implements and woman's work basket. *(Page 55)*
American Museum of Natural History.

3—Types of loom and weaving implements. *(Page 56)*
American Museum of Natural History.

4—Cylinders and stamps of terra cotta used in fabric or body printing.
American Museum of Natural History. *(Page 57)*

PLATE 7

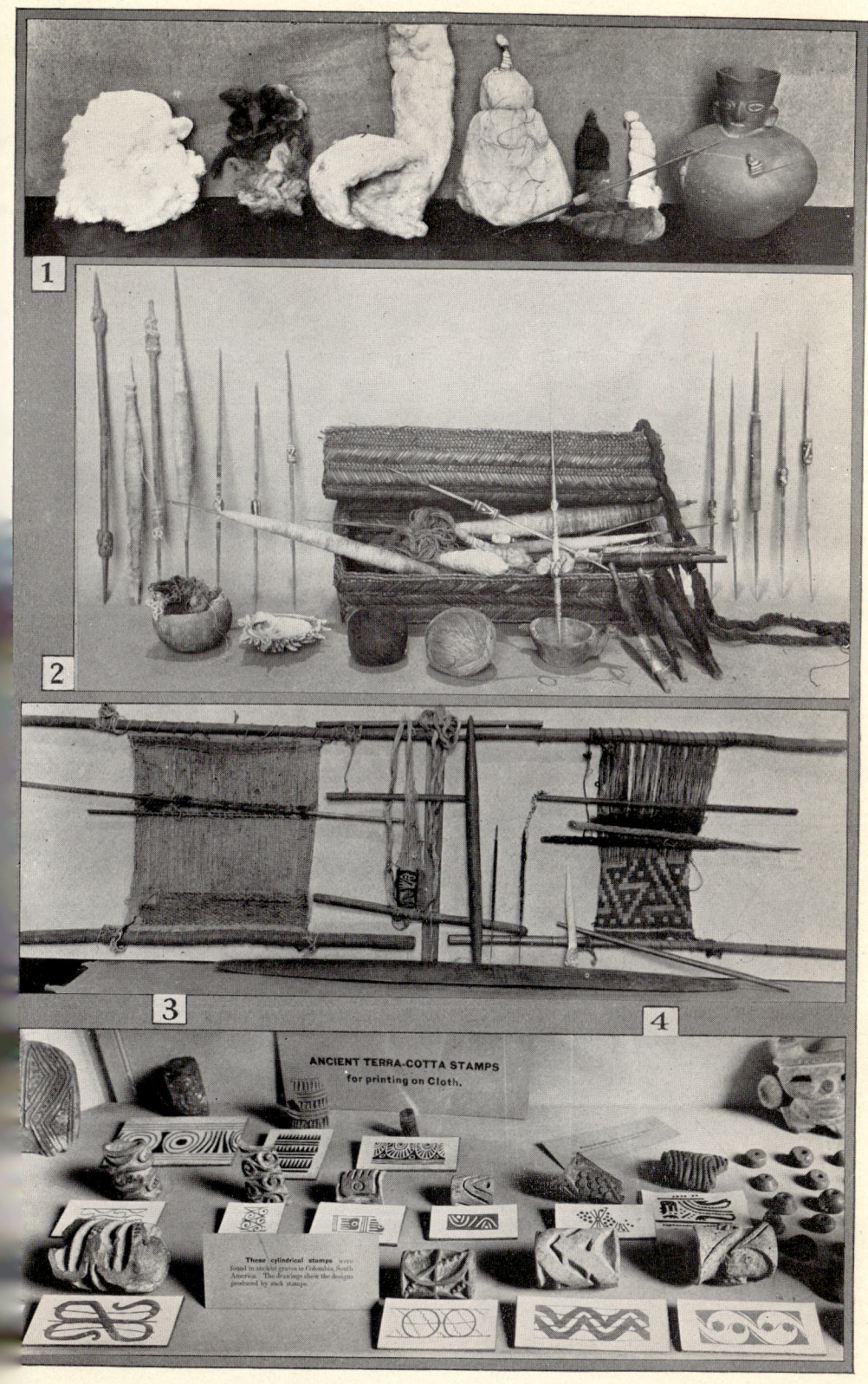

ANCIENT TERRA-COTTA STAMPS
for printing on Cloth.

These cylindrical stamps were found in ancient graves in Colombia, South America. The drawings show the designs produced by such stamps.

they met and conquered were themselves recent conquerors and despoilers of the older cultures. There nad evidently occurred, in these regions and in remote times, a series of dramatic national disasters, for which history can ascribe no certain reasons, and nations rich in culture, with immense material resources, high artistic and social attainments, had perished. Wattled huts had turned in the magic of man's proud genius into cities of carven stone. The rude council before the communal fire had broadened into the dignified senate and the shaman daubed in colored clay, festooned with the teeth and claws of animals, mumbling of the spirits of the dead, had been transformed into the magnificent priest, gorgeous in sacredotal robes, master and servant alike of a complicated ritual, all in the ordered sequence of time. And then at the zenith of power, in the flush of pride, had come the subtle change, and all that genius had devised, all that vanity had sanctioned, all that pride had willed vanished like some iridescent dream.

Ancient Peru was of these vanished people. From their graves have been taken some of the loveliest of woven webs. Here is found the complete catalogue of the weaver's art, traces of each fundamental subtlety of design and construction, and it is idle to assume that such maturity could have been reached, except over vast stretches of fruitful time. Here is in principle the perfect miniature history of the woven web. We could reconstruct each fabric and method of construction and decoration from the evidence of these tombs. We naturally find a different emphasis in technical processes than in Asia. In a general way, the Peruvians excel in the woven design, whereas the craftsmen of India more generally emphasize the printed and dyed fabrics. But all the principle techniques occur in both areas.

There is a certain snobbery even in the science of archæology. A careful distinction is drawn between the Old and the New World, a single tomb on the Nile, the contents of which in a general way are already well known, is opened with international complications, great and minute care as to each detail, particularly as these refer to newspaper copyrights and the division of spoils between the Egyptian Government and the more or less fortunate discoverer. All of these matters are supposed to advance the sacred cause of scholarship.

But the ancient empire of Peru, in whose sandy graves may lie the solution of one of history's most puzzling enigmas, where may repose the talisman to lead us surely and sanely backward into the shadowy cycles of those earlier civilizations, upon which all modern history, language, art and religions are built; all that we know of this has been gathered as a bi-product of the selfish quest of the mummy miners, seeking for precious metals during the three centuries of disorganized spoliation, since the Spanish Conquest.

The fortunate custom of burying with the dead their personal belongings, unfinished work and supplies of corn meal, dried fish and other foods, together with the narcotic coca leaf and the powdered lime, have given us a very accurate picture of their material culture. In the arts, these people achieved a great distinction. Their ceramics, in point of color and design, bear comparison with the finest examples of the potters art from any ancient or modern people. They were expert workers in gold, silver and bronze. They were not unskilled in the carving of precious and semi-precious stones, but it is in textiles that the chief ingenuity and artistry is shown. In the earliest graves we find cotton, llama, alpaca and vicuna hair-wools, human hair and a bast fiber the Agave Americana. But cotton was their principal fiber and its conversion into yarn and fabrics, dominates their technique.

There were, in ancient times, two varieties of cotton in Peru; one with a fine white lint of fair length, averaging from $1\frac{1}{4}$ to $1\frac{1}{2}$ inches, high in grade and excellent in character, the other a shorter, rougher, more uneven type of a reddish brown color. There is a shadow of evidence that this latter was an older type, since it shows less of the fine points of careful breeding. The brown cotton was said by Inca historians to have had some special traditional significance, probably due to it being the older form. On the other hand, the Peruvians, excellent dyers though they were in most colors, had some little difficulty with brown. The brown dyes were not as permanent apparently and had a deteriorating influence on the fiber. They appear to have been produced with some oxide of iron; this is of course merely conjecture, as no people dyed for eternity, but just as we do for the present moment. The fact that the older dyes have lived hundreds, perhaps thousands of years, could never have been a matter of importance to the actual dyers. The fact, however, that these great horticulturists kept the two types of cotton plants distinct and that this distinction exists today is proof enough that they saw in each type some special merit.

Fortunately it is possible to recreate from the actual fabrics and partially finished processes their complete technique.

The preparation of the fiber was simple, although it must be admitted very thorough. The seeds were first removed by hand. There is no record of any form of gin. The lint was next separated from the natural tangles and finally paralleled or carded, through pulling little bunches carefully apart and felting each little bunch of fiber into one large soft lap. This lap was about four inches wide and as long as the quantity of lint prepared demanded. It corresponds to a card lap in a modern mill.

This soft ribbon of carded cotton was then broken apart in short sections and formed into a bee hive cone, perhaps a foot high and six or eight inches in diameter at the bottom. It was from the larger end of the comb, that the fiber was drawn by the fingers in spinning, slightly twisted into a roving or partially spun thread and attached to the delicate little spindles of palm wood.

In the work baskets, found in all the women's graves, are usually a number of the tips of these combs, tightly wrapped in a cord made of the Agave Americana. These were evidently saved to be broken up again and formed into a large cone for spinning.

We need not rely on conjecture to describe their spinning processes. The Peruvians had the highly commendable custom of representing on their pottery every act of their lives. Certain types of Peruvian jars took the place in their culture the plastic arts assume in ours. One pottery vase, illustrated in this work, shows a woman in the act of spinning. The cone is held between one hand and the body and a thread, marked by a white line, crosses to the hand holding the spindle. This picture is of course generally accurate, yet it must not be forgotten that the Peruvian artist felt under no particular obligation to represent each detail of an act so common as that of spinning.

It is, therefore, not quite as clear today as we might wish it to be and we need a little imagination in reconstructing the actual movements. What probably happened was that the fibers were first drawn out partially wrapped around the middle of the spindle. The spindle was then rested in a shell or small wooden or pottery bowl, probably containing water, to moisten the fingers of the spinner. The cone was pressed lightly against the body and the left hand drew out more and more fibers, partially twisted them and the right hand operating the spindle at times increased the twist and wound on the finished yarn. Spinning was of course an intermittent operation and part of the time both hands were engaged in smoothing out the rough spots in the yarn, and putting in the requisite degree of twist. The true explanation of the exquisite perfection of their spinning is to be sought rather in the great skill of the craftsmen themselves than in the complexity of tools or processes.

Most Peruvian yarns in fabrics are two ply, some fine warps are three ply, and I have actually analyzed seven ply yarn. Hence they had need for a doubler spindle. The long spindles with carving at either end, were used for this purpose. Two single yarns were wound on the doubler, but not twisted together. The act of twisting and forming the plies was performed as the yarns were unwrapped and rewound on weaving bobbins. The carved ends prevented the yarn from slipping off and forming a tangle, just as the little beads of clay, metal and wood prevented the single yarns from slipping off the spindles. These meridional beads were not of the same technical significance as the whorl placed at the bottom of the spindle, and used in northern parts of South America and Mexico.

I have examined Peruvian single ply yarns, as fine as 200's in our cotton count. That is 200 times 840 yards to the pound. But there seems to have been little interest among these superb craftsmen for that ethereal lightness, which distinguished certain types of cotton spinning in India. Mere lightness of weight is, however, no absolute guide as to the skill in spinning. Our finer yarns have a greater tendency to irregularity than the coarser types. In the Peruvian yarns, no matter how fine they were spun in wool or in cotton, they are the absolute perfection of the spinner's art, indicating that had the wish been present, infinitely finer counts might have been spun.

The Peruvian loom has already been described in the previous chapter. Fortunately the burial customs of these peoples have preserved for us a few actual looms, thus removing all question as to their character. Some years ago, in preparing a scientific paper for the American Museum of Natural History, I found it necessary to reconstruct in imagination their processes of tapestry weaving in which art they excelled. Curiously enough, in all their wealth of pottery pictures, there was no known example at that time of weaving to guide me. It is a natural source of gratification, therefore, to write that a vase recently discovered by Dr. Julio C. Tello of Peru, not only justifies my conclusion, but leaves an accurate and delightful picture of this ancient and lovely craft.

The types of fabrics in Peru include every known construction as well as certain types, impossible to produce on machine looms. I have examined and dissected tapestries, double cloth, brocades, warp and weft stripes, leno, dobby patterns, crocheting and laces. Besides these types of design incorporated in a fabric,

PERU
PART 2

1—Vicuna wool tapestry with cotton warps showing rare plant forms.
American Museum of Natural History. *(Page 56)*

2—Peruvian mummy with false head and grave charms. *(Page 52)*
American Museum of Natural History.

3—Detail of Ica embroidery showing puma god with human heads.
American Museum of Natural History. *(Pages 56, 61)*

4—Detail of fine brown cotton voile with conventional motives.
American Museum of Natural History. *(Page 56)*

5—Rare example of Peruvian lace. *(Page 56)*
American Museum of Natural History.

6—Leno weave showing geometric patterns through crossing warps.
American Museum of Natural History. *(Page 56)*

7—Ica embroidery of puma god with human sacrifices. *(Page 56)*
American Museum of Natural History.

8—Double cloth in bird design. *(Page 56)*
American Museum of Natural History.

9—Leno cloth in bird design with tapestry border. *(Page 56)*
American Museum of Natural History.

10—Brocade pattern in fish convention. *(Page 56)*
American Museum of Natural History.

PLATE 8

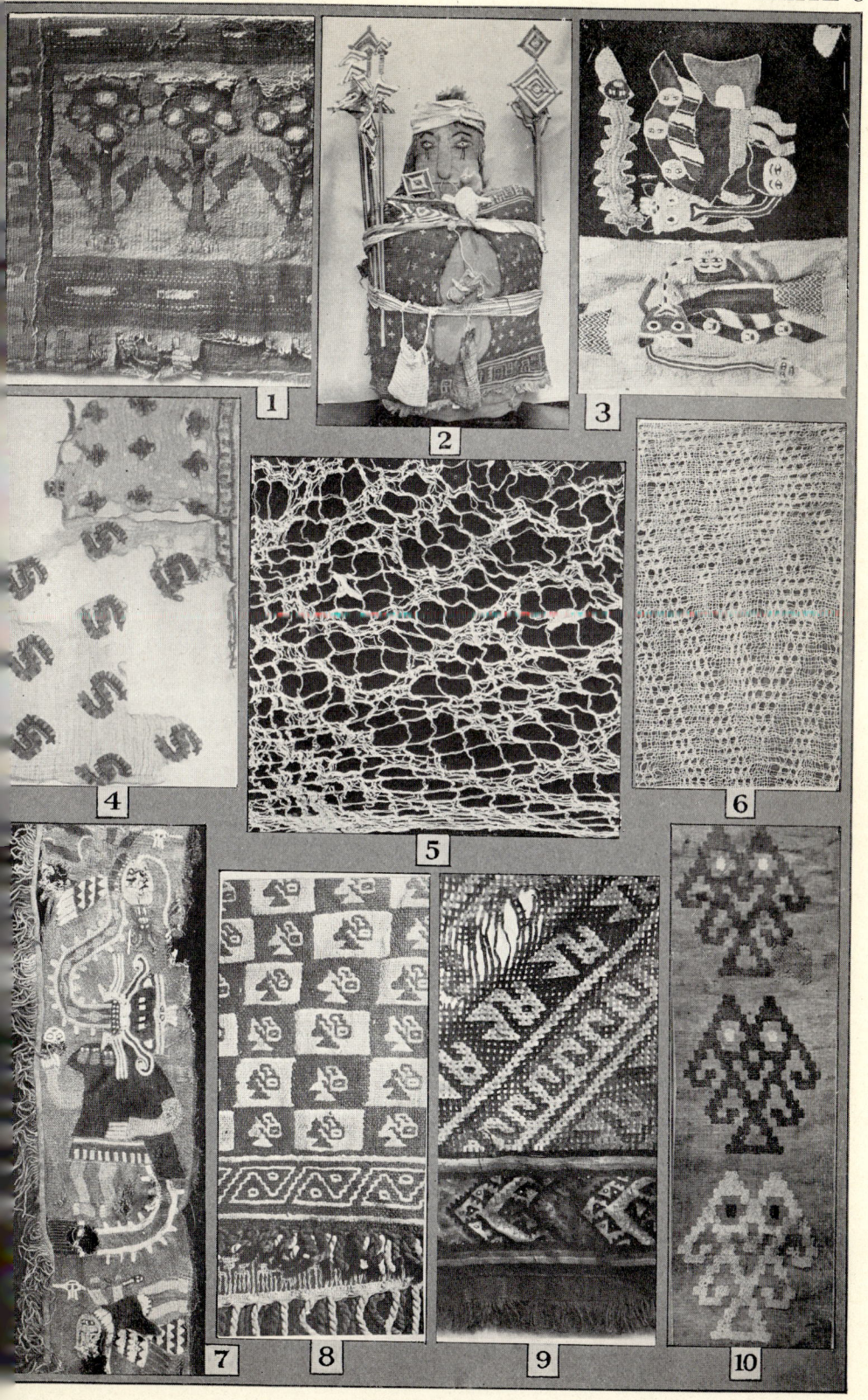

they were not lacking in knowledge of the application of ornament to the finished web. They were skillful embroiderers, delighting in techniques rather more subtle than any we know of from Europe or Asia, and more closely allied to weaving. Painting, or perhaps stamping on fabrics and even roller printing occur. At least we find small stamps and cylinders of terra cotta. These may, however, have been used in decoration of the human body. However, natives a century ago used to dig up the cylinders and print with them on cotton cloth.

The knowledge of weaving techniques and subtleties must have been universal, since even in the coarsest webs, we find the weaver passing easily and gracefully from one technique into another in the same loom piece. In all my experience with fabrics, I have never examined any collections which showed so perfect a sense of value of texture and color, or contain so few technical mistakes.

Their skill in using crepe yarns is truly amazing. One particularly fine voile of brown cotton, with interesting embroidered motifs, is a peculiarly lovely specimen. It is in the form of a kind of poncho or blouse, and since the Peruvians were a dark skinned people, this brown veil, with its brightly colored embroidered figures, must have been most effective. I have examined tapestries and brocades where between 200 and 300 weft yarns have been inserted to the inch in complicated patterns. Each of these designs had to be carefully calculated in advance and the least mistake would have upset the entire calculations and created a confusion in pattern. I have never seen an example of mis-weave in any of the finer Peruvian fabrics.

Naturally in such immense collections as have been

recovered from these graves, we find all degrees of skill and obviously some of the craftsmen were inexperienced. But as a general rule the teaching must have been sound and the pupils receptive. At the time of the Spaniards, there were great convents of women, who were compared by the historians to the vestal virgins of Rome. One of their chief functions was to produce for the Inca and the powerful nobles their beautiful garments of state, and it is from these perfect weavers, that we get the most exquisite textures.

In one of their forms of tapestry, at each change of color in the weft, the yarns are interlocked. This kind of weaving has no parallel that I am familiar with. They had a way, too, of outlining figures with black yarns made from human hair, and these yarns do not run at right angles as in most weaving, but in a more or less eccentric way, following the vagaries of patterns. Only an extremely skillful people could have taken such liberties with the basic principles of weaving.

One technique I have reserved until the last. In India and Java, in the Philippine Islands, in the Gobi Desert and in many other parts of the world through related techniques there is a process of applying design known as resist dyeing or more familiarly as batik. Herodotus in 450 B.C. describes this art among the natives of the Caspian Sea. It appears in the blue and white resist prints of Italy and Central Europe and is a part of the technical equipment of every first class fabric printing concern today.

The process in its simplest form is painting on a fabric with molten bees wax and then dipping the fabric in a dye bath, which does not dissolve the wax. The wax offers a resistance to the penetration of the color and when the wax is removed after the dye bath, design has been produced through the contrast between the undyed sections of the fabric and the dyed.

It is very difficult to say how the discovery of such a process could have been mere accident. It demands even in its simplest form an orderly sequence of processes, which must have been thought out in advance and deliberately performed with a fixed purpose in mind. To find such a process in Peru, evidently thousands of years old, was one of the most surprising and confusing discoveries I made in my research in Peruvian fabrics. To be sure in the pottery of Central America, occurs an analogous technique, known as lost color ware. Here a vessel is first dyed a rich red, then a design is painted on it in hot wax, using a chewed stick for a brush. The vessel is then immersed in a black dye and the wax removed in boiling water and consequently a design in black and red is created. It is of course possible, that the fabric arts here borrowed from the ceramic art, although in the natural order of development, the fabric arts are the elder. It is equally possible that this craft was transferred originally from fabrics to ceramics and the fabric arts lost in the moist climates, in which we find the pottery vessels. I do not know of any lost color ware in prehistoric Peru.

It is vain to attempt to describe the Peruvian palette of colors. Reds, purples and different shades of blue, green and brown predominate, but in such endless variety of tone and shade as to be impossible to limit by meaningless names. They have no comparison in our modern color scheme and belong in the same distinguished category as the colors of early Persia or of the reminiscence of the arts of Asia Minor found in the Egyptian tombs.

They were adept in all of the types of design which originated in the technique of weaving and their conventional forms are varied, well balanced and must represent not only a great time in development, but a high appreciation of the sense of values in balance.

It is possible in their conventions of animals to identify most of their complicated symbols. The cat, the bird, the fish and the human form predominate, although the llama and the monkey, the deer and even the serpent at times occur. We may see in these highly developed figures a reminiscence of the time when their art was based on propitiatory magic. All primitive peoples are more affected by animal than floral life and the Peruvians are no exception. Floral forms so persistent in our arts, since the Renaissance, have little place among the older peoples. Such floral forms as do occur, are probably in the nature of prayers for an abundant supply of water from the melting snow of the Andes and never occur without roots attached. It is easy to understand how a people, armed only with bronze spears and clubs and slings, had a great respect for a full grown jaguar or puma. Before society was fully organized, these cats must have been a serious menace to life, and no doubt rude, realistic pictures of them were made in many materials, in attempts to claim relationship with them and hence immunity from their destructive powers. The Peruvians were not in any sense a maritime people, still on frail rafts of rushes, they made journeys to the coastal islands for the guano used as fertilizer and they were fishermen as well. Consequently the booming surf of the Pacific, must have been a constant dread and the fish forms may represent propitiatory offerings to the malignant spirits of the deep. Certain it is that many fish forms

PERU

PART 3

Every technical method of applying ornament to woven texture is illustrated in the fabrics of pre-Inca Peru. They were not only master spinners, weavers and dyers but understood each subtlety of fabric construction. (Page 50)

1—Types of conventionalized fish design. C. W. Mead, "Peruvian Art." (Page 60)
American Museum of Natural History.

2—Brocaded cotton cloth with cat's head pattern. (Page 59)
American Museum of Natural History.

3—Very ancient poncho of vicuna wool with three-ply cotton warp. (Page 59)
American Museum of Natural History.

4—Unfinished fabric with brocaded designs. (Page 59)
American Museum of Natural History.

5—Types of conventionalized bird design. C. W. Mead, "Peruvian Art." (Page 60)
American Museum of Natural History.

6—Types of conventionalized cat design. C. W. Mead, "Peruvian Art." (Page 60)
American Museum of Natural History.

7—Pottery vase showing woman weaving, tied and dyed poncho. (Pages 58, 62)
American Museum of Natural History.

8—Double cloth bag used for coca leaf. (Page 60)
American Museum of Natural History.

9—Types of conventionalized human form design. C. W. Mead, "Peruvian Art." (Page 60)
American Museum of Natural History.

10—Brocade on leno weave. (Page 60)
American Museum of Natural History.

11—Double cloth with figures of men. (Page 60)
American Museum of Natural History.

12—Lace bag of Maguey fiber. (Page 58)
American Museum of Natural History.

13—Painted or stamped designs on cotton. (Page 61)
American Museum of Natural History.

PERU
PART 3

Every technical method of applying ornament to woven texture is illustrated in the fabrics of pre-Ican Peru. They were not only master spinners, weavers and dyers but understood each subtlety of fabric construction. *(Page 56)*

1—Types of conventionalized fish design, C. W. Mead, "Peruvian Art."
American Museum of Natural History. *(Page 60)*

2—Brocaded cotton cloth with cat's head pattern. *(Page 56)*
American Museum of Natural History.

3—Very ancient poncho of vicuna wool with three-ply cotton warp.
American Museum of Natural History. *(Page 56)*

4—Unfinished fabric with brocaded designs. *(Page 56)*
American Museum of Natural History.

5—Types of conventionalized bird design, C. W. Mead, "Peruvian Art."
American Museum of Natural History. *(Page 60)*

6—Types of conventionalized cat design, C. W. Mead, "Peruvian Art."
American Museum of Natural History. *(Page 60)*

7—Pottery vase showing woman wearing tied and dyed poncho.
American Museum of Natural History. *(Pages 58, 59)*

8—Double cloth bag used for coca leaf. *(Page 56)*
American Museum of Natural History.

9—Types of conventionalized human form design, C. W. Mead, "Peruvian Art." *(Page 60)*
American Museum of Natural History.

10—Brocade on leno weave. *(Page 56)*
American Museum of Natural History.

11—Double cloth with figures of men. *(Page 60)*
American Museum of Natural History.

12—Lace bag of Maguey fiber. *(Page 56)*
American Museum of Natural History.

13—Painted or stamped designs on cotton. *(Page 57)*
American Museum of Natural History.

PLATE 9

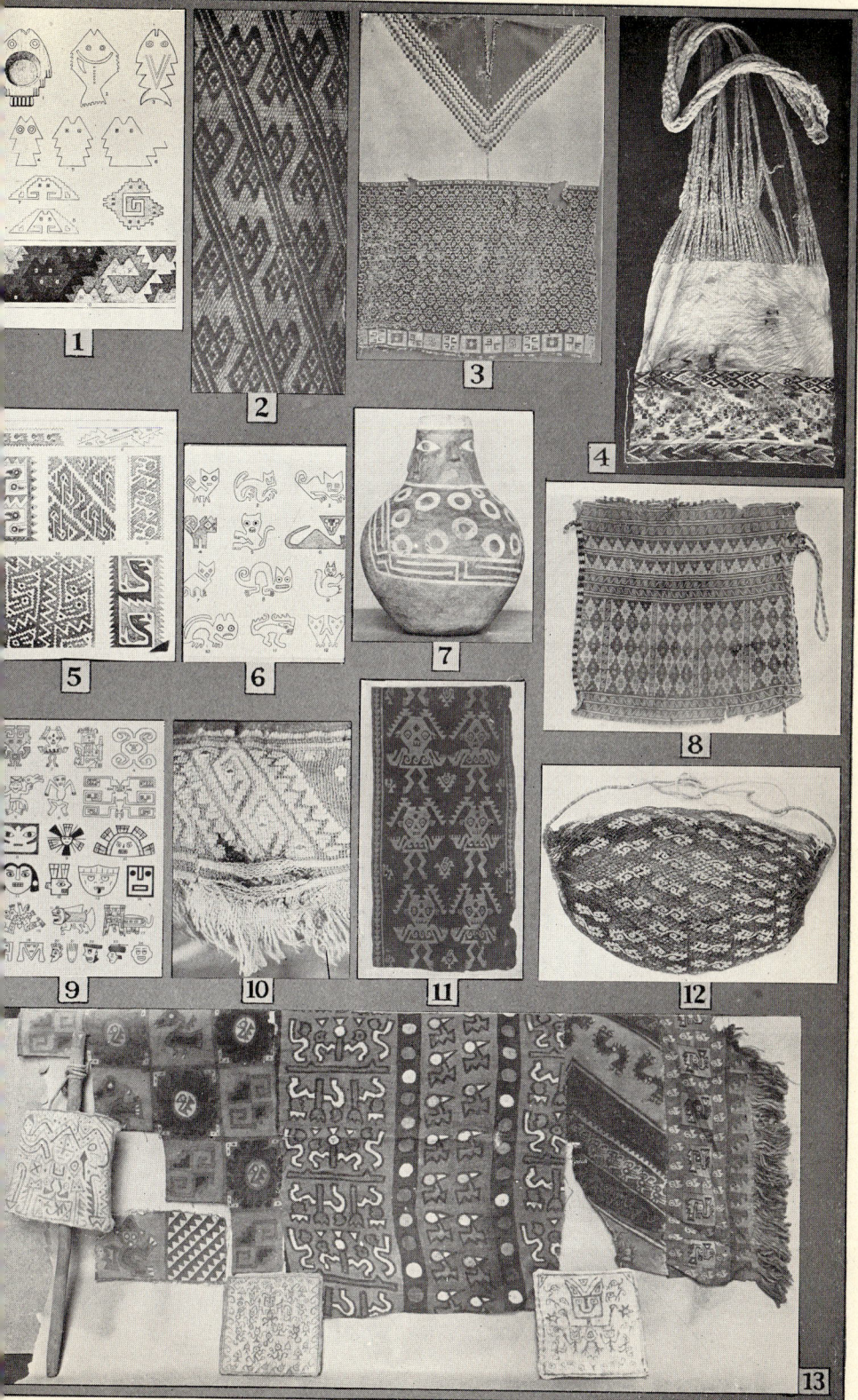

made from thin silver have been found on the islands under thirty feet of guano. The bird had perhaps a double significance. A desert people, living in the open spaces on the coast, were the victims of great and terrific wind storms and the bird may have been, therefore, the symbol of the air. On the other hand, the returning birds, or rather the restoration of the mating plumage of the birds, may have had some association in their minds with the melting snows on the ice caps of the Andes and the water supply that brought fertility and color to their lives. The human form is seldom shown except as an armed man, in many cases bearing the gruesome trophies of battle, the severed heads of his enemies; and here we have an expression of the fear of warfare as an incentive to design.

When, however, a people vary their symbolical forms and fetish ornaments and develop them into a highly complicated rythmical composition, it is a fair assumption that they have an æsthetic rather than a sacerdotal significance. I have in mind one particular piece of embroidery from an old Inca shawl, the border of which shows the panther god decorated with human heads, as the symbol of human sacrifice. Each figure is in rythmical repetition to the succeeding and each in a distinct color combination, yet all in perfect harmony. If the puma god really held any terrors for the artist that created this design, it must be admitted that in the interest in composition and color, he had taken great liberties with the object of his fear and worship.

CHAPTER VI

INDIA

LIVING and classical languages alike indicate India as the original source of cotton. Madras, chintz, calico, lawn, muslin and mull are a few of the words, Indian in origin, we commonly use to designate cotton fabrics.

The first word we may clearly define as meaning cotton is the Sanskrit, "Karpasi." The centuries have changed this word but little into "Kapas" in modern Hindustani. In the writings of the Hebrew scholar Josephus appears the word "Chethon," but this is rather doubtful, since it may refer to linen or ramie as well as cotton. Around the word "Othonium," a fabric dealt in by the Phœnicians, is an equal uncertainty. There can be no question, however, that "Karpas" in the Book of Esther, means cotton and was derived from the Sanskrit original.

The early Greek writers simply refer to cotton as the fiber of wool-bearing trees and give it no name. Herodotus, in 425 B.C., writes, "The wild trees in that country (India) bear for their fruit a fleece surpassing those of sheep in beauty and quality and the natives clothe themselves in cloth made therefrom."

Nearchos, one of Alexander the Great's Admirals, in 300 B.C. mentions these trees and accurately describes the costumes of the people. "The natives made linen

INDIA

The distribution of resist dyeing is very wide both in point of time and geographically. It is principally known as batik, bandhani, tie and dye, and elastic painting. Mallet beeswax, applied either with a brush or other implement, is generally used, although farinaceous starches and clays are at times applied. In tie-dyeing a cord or thread is used.

This page of illustrations is intended to convey some idea of how widespread this technique is. Herodotus mentions it in 431 BC among the natives along the borders of the Caspian Sea. So far as other portions of Asia are concerned, including the Malaic Islands, Africa and Europe, this art evidently spread from southern India.

The appearance of this technique in pre-Columbian Peru is difficult to explain and may be an accidental dual discovery. It is, however, apparently a very ancient art.

1—*Sandinia* from A. Von Le Coq's "Chotscho." This is ascribed to the 7th-9th Century of the Christian Era.   (Page 69)

2—Twelfth Century Sumatran Cloret Hanging.   (Page 64)
Kevorkian Collection.

3—Prehistoric Peru.   (Page 68)
American Museum of Natural History.

4—Oldest Batik in Japan credited to the Twelfth Century   (Page 70)
Shōsōin at Nara in Japan.

5—Egyptian wax painted silk 7th, 8th or Ninth Century.   (Page 72)
British Museum.

6—Modern tied and dyed silk from India.   (Page 70)
Brooklyn Museum.

7—Modern elastic block print from Schleswig-Holstein.   (Page 70)
Brooklyn Museum.

8—First waxing process of Javanese Batik.   (Page 70)
Collection of the Author.

9—Fabric in progress of tie-dyeing called caubies, prehistoric Peru.   (Page 68)
American Museum of Natural History.

10—Examples of tied and dyed fabric from modern Bombay.   (Page 72)
Collection of the Author.

11—Bedspread from Philippine Islands, showing process of tie-   (Page 70)
dyeing and resist printing.
American Museum of Natural History.

INDIA

The distribution of resist dyeing is very wide both in point of time and geographically. It is variously known as batik, bandhani, tie and dye, and mastic printing. Molten bees' wax, applied either with a brush or other implement, is generally used although farinaceous starches and clays are at times applied. In tie-dyeing a cord or thread is used.

This page of illustrations is intended to convey some idea of how widespread this technique is. Herodotus mentions it in 450 B.C. among the natives along the borders of the Caspian Sea. So far as other portions of Asia are concerned, including the Asiatic Islands, Africa and Europe, this art evidently spread from southern India.

The appearance of this technique in pre-Columbian Peru is difficult to explain and may be an accidental dual discovery. It is, however, apparently a very ancient art.

1—Specimen from A. Von le Coq's "Chostscho." This is ascribed to the Fifth Century of the Christian Era. *(Page 69)*

2—Twelfth Century Armenian Church Hanging. *(Page 64)*
Kevorkian Collection.

3—Prehistoric Peru. *(Page 58)*
American Museum of Natural History.

4—Oldest batik in Japan credited to the Twelfth Century. *(Page 76)*
Museum at Nara in Japan.

5—Egyptian wax painted silk from Eighth or Ninth Century. *(Page 75)*
British Museum.

6—Modern tied and dyed silk from India. *(Page 76)*
Brooklyn Museum.

7—Modern mastic block print from Schleswig-Holstein. *(Page 76)*
Brooklyn Museum.

8—First waxing process of Javanese Batik. *(Page 75)*
Collection of the Author.

9—Fabric in process of tie-dyeing rolled on bias, prehistoric Peru.
American Museum of Natural History. *(Page 58)*

10—Examples of tied and dyed fabric from modern Bombay. *(Page 79)*
Collection of the Author.

11—Bogobo headdress from Philippine Islands, showing process of tie-dyeing and finished turbans. *(Page 76)*
American Museum of Natural History.

PLATE 10

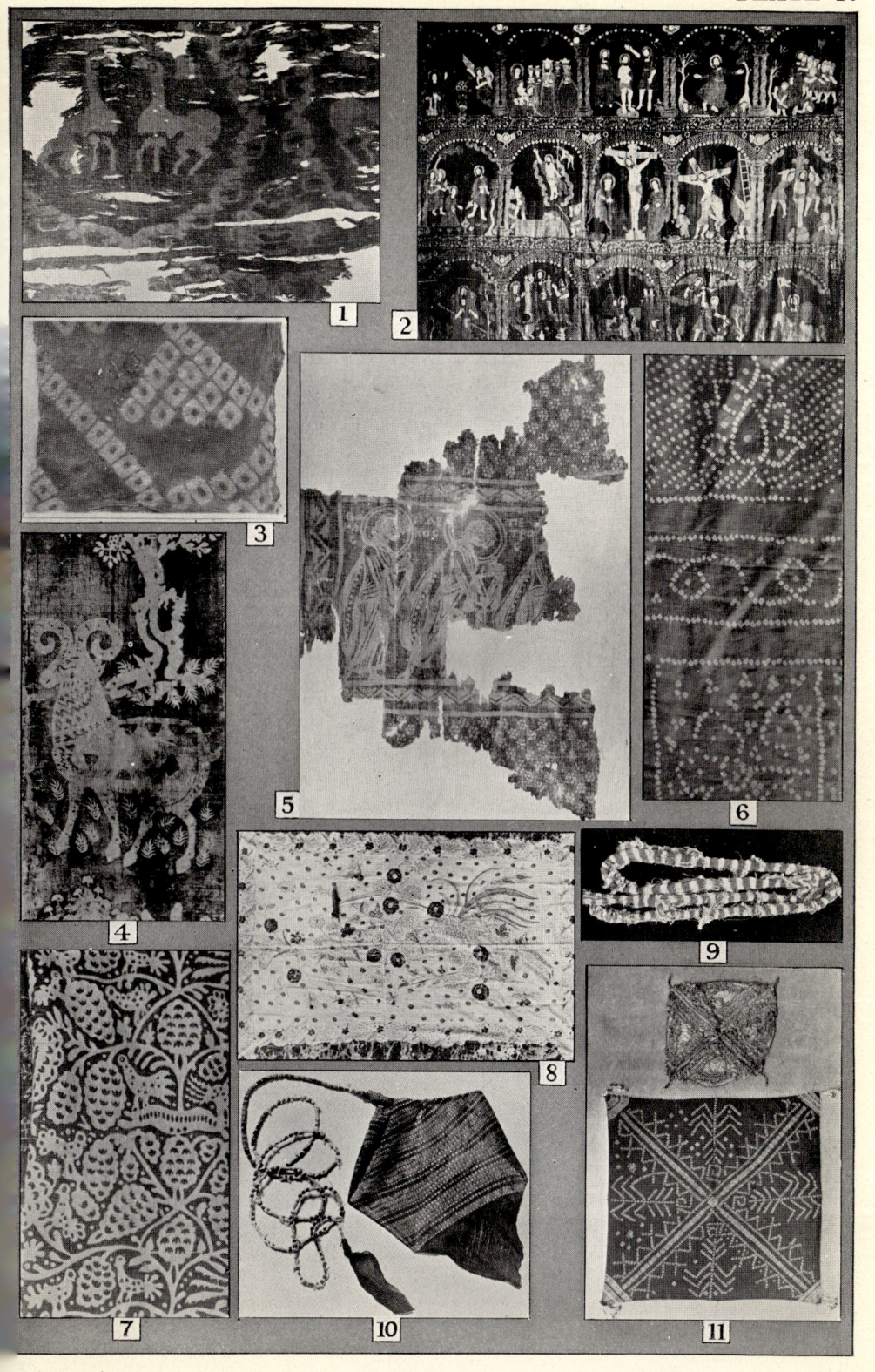

(cotton) garments, wearing a shirt which reached to the middle of the leg, a sheet folded over the shoulders and a turban around the head."

In a fragmentary Greek drama of 169 B.C., appears the word "Carbasina" referring to this tree wool, proving that at this time the Sanskrit word had already become familiar in the Mediterranean area. In the digests of Justinian, appear two words, "Carbasa" meaning cotton yarn and "Carbasum" meaning cotton fabrics, both evidently derived from the Greek adaptation of the older Sanskrit word. In 70 A.D., Pliny mentions tents of "Carbasus."

Thus briefly across classical times, we can trace the philology of the word and a common recognition of India as the source of supply.

With the exception of Moorish Spain, Europe of the Middle Ages knew little of cotton. One of the earliest names we encounter is "Bombassium," derived from "Pambax," meaning tree wool, or maybe a confusion with an earlier name for silk. "Barometz," referring to the mythical Scythian sheep, was also used in Central Europe about this time.

The modern word "cotton" is derived from the Arabic "Kotn," appearing in the Medieval Latin form of "Cotonum," in the ledger of an Italian merchant of the 13th century. The Spanish word "Algodon," is obviously a corruption of the Arabic "Kotn." In Italian the word becomes "Coton," in French "Coton," in German "Kattun," in Russian "Kotnja," and in Roumanian "Kutnie."

These different names of fiber, prove that if the classical world knew it as an Indian product, Medieval Europe regarded it as either from the Near East or Arabic in origin.

This does not prove conclusively, however, that the techniques and decoration we so generally associate with cotton had the same origin. As a matter of fact, most of the woven designs probably originated in Asia Minor and are a reminiscence of the Assyrian, Grecian and Parthian arts.

The Chinese explorers in the early centuries of the Christian Era are profuse in their praise of the skill of the Parthian embroiderers in metallic threads. The Chinese at this time were not skillful pattern weavers themselves, but were marvelous embroiderers and they probably described as embroidery what were actually woven textiles.

All evidence, however, points to the Indian Peninsula as the original home of painting, printing, resist and mordant dyeing, on cotton fabrics.

Herodotus in 450 B.C., writing of the peoples on the borders of the Caspian Sea, says:

"They have trees whose leaves possess a most singular property. They beat them to a powder and then steep them in water. This forms a dye with which they paint figures of animals on a garment. The impression is so strong, that it cannot be washed out and it appears to be interwoven in the cloth and wears as long as the garment."

There can be little question that the father of history is here describing the process of resist dyeing with indigo and beeswax, which of course would appear to him like direct painting on the cloth.

The Greek physician Ktesias in 400 B.C., mentions the flowered cottons emblazoned with glowing colors, much coveted by the fair Persian women and exported from India. The Greek Megasthenes in 300 B.C., an Ambassador to the Court of Chandragupta, who spent many years in India, writes:

"In contrast to the general simplicity of their lives, the Indians love finery and ornament. Their robes are worked in gold and ornamented with precious stones, and they wear also flowered garments made of the finest muslins."

Pliny the elder, in 70 A.D., describes a process of mordant dyeing fabrics in Egypt, which reminds one very much of the actual process in India described by a French Jesuit in the 18th century.

The art of resist dyeing spread among all the peoples who came in contact directly or indirectly with Indian influence, and there is still a reminiscence of this craft among the peasants of Europe. These facts seem to establish India as the home, not only of cotton, but of certain of the processes of dyeing and printing cotton.

In the Statutes of Manu 800 B.C., there is a brief but interesting passage laying down certain rules of guidance for the guild of cotton weavers.

"Let a weaver who has received ten palas of cotton threads, give them back increased to eleven through rice water, which is used in the weaving. Whosoever does otherwise, shall pay a fine of 12 panas."

This terse admonition is of more significance to the technician, perhaps, than to the historian. In the first place it proves that the crafts of cotton weaving and spinning were distinct. In the earlier phases of the primitive textiles, weaver and spinner are the same individual and the entire craft is an occasional domestic occupation. This passage indicates that cotton yarns were spun in the homes, gathered by merchants and sent to the weaver guilds to be woven into fabrics. Here is, therefore, not only a specialization in industry, but the intrusion of modern merchandising ideas. Merchants controlled, no doubt, the activities of the craftsmen and sold the finished product at a profit.

The addition of rice starch to the warp threads was not necessarily an adulteration. Cotton warps, in even the simplest kind of a loom, where the shed is made with heddles, should be prepared with a size so as to slip easily through the loops in the heddle frame. No starch or size is necessary in the gentler weaving of the true hand loom. I think this proves, therefore, that at least as early as 800 B.C., the type of loom used in India in the 16th century and still found in certain favored regions was already in existence.

Cotton is also mentioned in this venerable work, as the fiber from which was spun the tri-parte caste thread of the Brahmans. It is not conceivable that a powerful and ancient religious body would choose any material for such a purpose, unless it were hallowed by long established tradition. It is worthy of notice, as well, that both these references to cotton treat it as a matter of common knowledge, a well established, familiar, economic and artistic fact, not as some recent cultural intrusion.

One of the most romantic phases of the cotton story in India concerns the gossamer muslins for which Dacca was once famous. There are occasional references to these fabrics among the classical writers, but a surer proof exists in the Indo-Greco statuary of the first and second centuries of the Christian Era. The marble images of Buddha, so evidently inspired by Athenian statues, show a drapery quite distinct. One significant feature of these statues is the way in which a fabric of incredible lightness has been perfectly draped in natural folds on the human form. No artist could model such a quality, unless familiar with it.

To reduce the lightness of Dacca muslin to a modern formula is not difficult. There is an apparently accurate record in the 17th century of fabrics fifteen yards long and one yard wide, of such incredible lightness as to weigh only 900 grains or 60 grains to the square yard. This would mean about 73 yards to the pound. The lightest fabric made by Swiss hand loom weavers averages between $16\frac{1}{2}$ and 17 yards to the pound, 40 inches in width, or roughly four or five times heavier.

It was natural that a highly imaginative and artistic people should give to these muslins poetic names, such as the "Evening Dew," Shahnnam; "Running Water," Abrawan; "Mull of Kings," Mull Mul Khas; "Presentation," Shaugatai; or "Sweet Like a Sherbet," Sharbatti.

It is doubtful if any of the finer grades of Dacca muslins ever became objects of trade. Even in the days of cheap labor they were very expensive and were, like certain of the finer rugs of Persia, reserved for royal usage and as gifts to friendly courts.

There are many charming legends about these misty veils, some no doubt as great a tribute to the imagination of the writers as to the skill of the spinners or weavers. Still we must admit that proof enough exists to establish the muslins of Dacca as the most delicate cotton fabrics ever fashioned in a loom.

It is said that in Dacca a special cotton was grown

of a very fine and long fiber, but this has largely if not entirely disappeared. It is interesting to note that the spinning wheel, which at a very early date became general all through India, was never used to spin these yarns. The method of spinning was like that described in prehistoric Peru. The point of the spindle (a fine needle of bamboo, with a little pellet of clay at the bottom) rested in a dish, containing water, to moisten the fingers of the spinner. The act of spinning was performed by twirling this little sliver of bamboo and working the fiber between the fingers. It was a matter of exquisite training, not of complex implements or processes. You can no more explain such spinning than you can describe how Fritz Kreisler plays a violin.

There were many types of cotton fabrics all through India. Brocades, embroideries, heavy fabrics for winter use, textures of metal and silk threads and other refinements of weaving were common. Each city had apparently some particular phase of this craft in which it excelled.

The rise of cotton in early international trade becomes highly visible with the great Mohammedan conquests of the Eighth Century. The flight from Mecca to Medina, from which the Mohammedan calendar is reckoned, was in 622 A.D., the first invasion of Egypt in 639 A.D., the Mediterranean coast of Africa in 703, India in 711, Spain in 712 and Syria in 737. In the brief span of a generation, these vigorous peoples overran practically all of the ancient classical civilization. The extent of their domain was finally determined by the stubborn resistance of the Castillian knights and the dauntless lances of Hungary. There was a time when the sons of the Prophet stormed at the gates of Vienna and measured their valor against the iron chivalry of France.

The picture is by no means all shadow, for to the Moors and the Saracens we owe no small debt for the preservation of the priceless remnants of classical culture. After the fiery warrior came the able and tolerant administrator, the patient scholar and the inspired craftsman, garnering in ancient fields the golden sheaves of learning and beauty. Great as were the teachers, it must be admitted that the pupils were not unworthy.

More than this, the Arab merchants, opened the old avenues of trade, established new caravan routes and linked together many peoples in the ancient world through the first truly comprehensive international commerce. It was through this medium that the arts spread, that the arts of ancient peoples became enriched through intrusion and adaptation and restored to vigorous life.

There can be no question that the pre-Mohammedan peoples of India were skilled artists in cotton fabrics. The above quoted writings of classical authors attest this fact beyond dispute. In the wall paintings of the Ajanta Caves of the Fifth to the Seventh Centuries, there are innumerable representations of cotton costumes and wall hangings of great distinction. The fabrics found in the frozen sands of Gobi, while not often of this fiber, still attest a high artistry in textiles and are obviously related to the famous wall paintings. At the same time the arts of cotton, which first attracted European notice and which so powerfully affected the artistic and commercial life of Europe in the Sixteenth and Seventeenth Centuries, were unquestionably the result of Mohammedan influence and creation.

It may be well to recall that these conquerors of India had come from the greatest centre of art and culture in history. They were familiar with the arts of the late Greek Empire, the ancient arts of Asia Minor and Parthia. Undoubtedly they made a deep impression on the native crafts and created a complete change in the spirit of design almost at once. Their influence in a sense parallels that of Spain on the primitive arts of the New World.

There are very few actual specimens of cotton cloth from India we can safely date before the Sixteenth Century. The largest collection of early cotton, perhaps, is now in the Brooklyn Museum and was discovered in the ruined city of Amber, deserted in the Sixteenth Century on the word of a soothsayer and made famous in Kipling's story the *Naulahka*. These priceless records have been carefully restored and mounted and form one of the most beautiful and interesting exhibits of this delightful museum. These fabrics were wall hangings representing pictures of the famous British ambassador, Sir Thomas Roe (1615–1618), dressed in the costume of James the First's time. Other panels are decorated with pictures of wild tribes of India, resembling faintly the fanciful pictures of our own North American Indians. Another is a beautiful conventionalized drawing of a plant with large purple blossoms, not unlike the sacred cotton tree of India, which also has purple blossoms.

In color as in the freedom of design, these fabrics are far superior to any of the later Indian cottons made for the European markets. It was the custom then as later to hang these illustrated curtains in the rooms of palaces, wherein distinguished guests were entertained. While excellent, the construction of the cloth is not particularly fine. It has been said that curtains of this character were often changed, the idea being to represent on the wall scenes of interest, and to recreate indoors the spirit of the outer world. The more familiar patterns of the Seventeenth and Eighteenth Centuries often contain the beautiful form of the doorway of Taj Mahal, with a sprinkling of flowers and blossoms in the centre. It is as though one looked through a doorway into the world of outer beauty.

European costumes very early strongly influence the fashions of the East and introduced Occidental notions of tailoring. But the ancient costumes of India were largely if not entirely draped, uncut fabrics, as we might expect among a textile people living in a tropical country. The long shirt or tunic described by the Admiral Nearchos in 300 A.D., the loin cloth, the shawl or sari, the kilt and the turban were the principal articles of attire. Ornament was chiefly on the ends or along the selvedge of the fabrics, and consisted of woven designs in gold and silver threads or embroidered patterns. Some of the smaller figures used in the ground patterns of chintzes occasionally appear, but with this exception; the costumes were very simple in design, compared to the gorgeous wall hangings, this being, of course, in keeping with their good taste and the sense of the appropriate.

It must be frankly admitted that our actual knowledge of their textile arts, up to the Sixteenth Century, is extremely limited. Fabrics are, no doubt, the commonest forms of art and have the widest usage. At the same time, they are the most perishable. Few places in India have the peculiar climatic and natural conditions that have made possible the preservation of fabrics of Egypt and Peru and thus place these records of skill beyond cavil. Perhaps the surest measure of their artistry is to be found in the great influence they exerted on Europe.

Europe, from at least the time of Alexander the Great, has traded with the Orient. There are meager records of commerce in light woolen cloths, in ornamental glassware, black lead, coral and wine from Europe, and of spices, ivory, gems, silk and cotton fabrics from the East. In the Second Century of the Christian Era, Claudius Ptolemy, the great calligrapher, who so powerfully influenced the scientific thought which directed the navigators of the Fifteenth Century, has left us a map of Asia and Mediterranean Europe, that is, all things considered, amazingly accurate. This map, with others of almost equal interest, appears in John Fiske's *Discovery of America*. With the exception of certain details added by Marco Polo in the Thirteenth Century, this map represented the high water mark of Europe's knowledge of the geography of the East, down to the famous voyage of Vasco da Gama at the close of the Fifteenth Century.

This absence of detailed, geographical information from the East, does not mean that intercourse had ceased between the times of Ptolemy and Messer Marco, but that scientific observation of fact was submerged either by theological dogma regarding the shape of the earth, or the indifference of the trader. From the Fourth Century on, the schismatic sect of the Nestorian Christians penetrated the East, establishing monasteries in Herat and Kashgar and even in far China.

During the dark Eighth, Ninth and Tenth Centuries, and later, Constantinople, the seat of the Eastern Roman Empire, was the bulwark of benighted

Christian Europe against the fierce attack of the rising Mohammedan powers. The rise of the fanatical Turkish power, overthrowing the tolerant and learned Arabs, was answered in Europe by the sudden ardor of the Crusades. There can be no question that this contact between vigorous Europe and the culture of the Near East, particularly in Constantinople, was a powerful stimulant to the revival of art and learning in Europe.

Here begins the rise of the great trading cities of Venice and Genoa, and their bitter struggle to monopolize the eastern markets. During this period, the open-minded policy of the great Mogul conquerors of Asia permitted European travelers, both lay and cleric, to visit the East, and among these was the delightful Marco Polo.

Late in the Thirteenth Century he reached the Coromandel Coast and made pointed mention of the beautiful muslins and colored chintzes of Masuliputam. In 1375, appeared the Catalan map showing for the first time the Pacific Ocean and the Islands of the Indian Sea, including Java.

This map powerfully influenced the great Toscanelli, the scientific advisor of Columbus, in establishing his theory of a spherical globe.

The fall of the liberal Mogul Empire and the rise of the Ming Dynasty, in 1368, closed Asia to European exploration, and was perhaps the beginning of the great urge to find an all water route to these rich markets.

But a still more terrible danger threatened, for the Turks, crushed by the Crusades, arose again in 1365, entered the Balkan Peninsula and, finally, in 1453, captured Constantinople and thus cut off the Genoese trade. Later spreading into Syria and Egypt, they effectually throttled the prosperous commerce of Venice. Such illicit trade as remained was seriously hampered by the teeming corsairs of the Mediterranean who were not subdued until in the beginning of the Nineteenth Century by the infant navy of the United States.

I have been at some length in describing the early trade between Europe and Asia, because of a certain tendency to belittle the influence of this contact among modern historians who see in the dominance of Asiatic arts, perhaps, some reflection on the creative powers of early Europe and the great economic contributions of the New World to later Europe. I hope I have made it clear that such of our earlier arts, implements and processes, as indicate Asiatic intrusion, had ample time to have acquired these traits.

In the middle of the Fifteenth Century, begins that remarkable series of voyages of trade and discovery, which ended in 1492 with the Discovery of America by Columbus and the passage of the Cape of Good Hope by Vasco da Gama, a Portugese captain, and his successful voyage across the Indian Ocean to the city later known as Calcutta in 1497.

This voyage effectually opened the long desired markets of India to Europe and created more interest at the time than the Discovery of Columbus, who believed he had found the Orient, but at a less populous and wealthy point than his Portugese rival. Cotton goods were brought back to Lisbon by Vasco da Gama and while, perhaps, spices and other more precious material were of greater importance, still these fabrics attracted attention, since in 1498 Obvarado Barbosa, a trade adventurer, who followed da Gama in the next year, mentions painted cotton cloths or "Pintado," a name long retained, referring to calico.

From here on the history of cotton and cotton fabrics belong to Europe so far as our interests are concerned. Almost from the inception of this trade, European ideas in designs and quality affected and debased the ancient arts. Portuguese, Spanish, Dutch, English and French navigators, merchants and adventurers, in the order named, entered the East and established colonies and factories to develop the trade and often to exploit the natives.

The remainder of this chapter I shall devote to a brief description of the techniques of dyeing and printing calicoes in India, reserving further comments on this amazing trade for the following chapter.

In G. P. Baker's *Calico Painting and Printing in the East Indies*, appears a letter, written by Father Coeurdoux, a Jesuit Missionary at Pondicherry in January, 1742, describing the painting and dyeing of the finest cottons. Curiously enough, this long and more or less technical description corresponds vaguely with the brief hearsay account Pliny gives in 70 A.D. of mordant dyeing in Egypt.

Briefly there were four principal methods of decorating cotton fabrics: resist dyeing, mordanting, stamping and painting. The earliest was resist dyeing; that is, painting the cloth with molten wax or clay and then dyeing. Where the resist substances adhered to the fabric, the dye could not penetrate, and by applying wax and dyeing a number of times, inconceivably intricate patterns could be produced. In India the tools for doing this were little bent forks of iron, the points of which were joined together on the principle of a stylographic pen, and wrapped with woolen yarn or hair to hold the dye, wax or mordant as the case might be.

In Java, a little vessel of copper with a delicate spout of copper leaf attached to a reed handle was used. The opening of the spout was so small that the melted wax could only come out through capillary attraction. This made it possible to draw patterns of the greatest delicacy.

Another process, known as tie dyeing, is also a form of resist. Little bunches of the fabrics are caught up and tied with a thread that resists the action of the dye and the exquisite degree of skill in this simple method is beyond praise. This art was practiced not only in India, but reached as far as the Philippine Islands and into Central Europe, across the Balkans, and even into Thibet and Japan.

Mordant dyeing consists of applying, with a brush or stamp, different chemical substances to a fabric, which cause the fabric when placed in a dye vat to take on different colors. In other words, certain substances have distinct color affinities, and by regulating these, patterns can be produced. This is, however, an extremely limited process and requires a broad knowledge of chemistry. It does not appear in the later calicoes and painted cloths to any extent, although from the early description, particularly that of the French Jesuit, it must at one time have been an important part of the older processes.

At a very early date, carved wooden blocks or wood with little ribbons and pegs of metal inserted were used for designs which had become classical and for which there was always a ready market. In addition to this, there was brush work and the use of little pads of cotton dipped in the dye.

We can trace the arts of resist dyeing all through the Islands of the Indian Ocean and even into Thibet,

INDIA

1—Eighteenth Century printed and painted Indian wall hanging.
   Metropolitan Museum of Art. (*Page 70*)
2—Eighteenth Century printed and painted Indian wall hanging.
   Metropolitan Museum of Art. (*Page 71*)
3—Detail of Javanese batik scarf, Nineteenth Century. (*Page 76*)
   Collection of the Author.
4—Wall painting showing cotton design from the Ajanta cave between the First and Fifth Centuries A.D. (*Page 69*)
   The Paintings in the Buddhist Cave Temples of "Ajanta"—John Griffiths.
5—Modern tied and dyed Indian scarf. (*Page 80*)
   Collection of the Author.
6—Detail of Javanese batik scarf, late Eighteenth Century. (*Page 76*)
   Collection of the Author.

PLATE 11

Japan and China. The Moros carried it to the Philippines and the turban of the head-hunting Bogobo Tribe is a little square of reddish brown cotton, covered with the circular marks characteristic of this craft. In Japan, the arts were first introduced by Dutch traders in Nagasaki. An old Japanese book, published in 1720, shows a merchant of Holland painting a design with a brush, and the same book contains many patterns that are safely Indian. The oldest Japanese designs all have the true character of resist, and most of their earlier stencil patterns were used with a paste made from wild rice, not with direct dyes.

There is more sentimentality than truth in the idea of lost arts. A long study of craft history has led me to doubt the mortality of any medium to create beauty, once it is firmly established. Arts become atrophied, diminishing to a vanishing point, through many causes, but somewhere will abide the little spark ready to kindle at a moment's notice. The familiar human tragedies of history, have relatively little influence on these things. Wherever an audience appears with the power of appreciation, there will be found both the skill to express and the power to create and revive beauty. This is peculiarly true of the ancient arts of cotton.

For a century, the mechanical products of the Occident in their worst form have deluged the Orient. Every banality of design, each crudity and inanity of color, every debasement of structure, that could be conceived in the sordid soul of cheapness, have literally been dumped in these ancient homes of loveliness. It would seem that the spark must have been smothered and many earnest men and women, myself modestly included, have believed and written that the damage was irreparable. On the other hand, many smug and sanctimonious individuals have striven to prove the immense benefits conferred by machinery and trade on these devoted peoples, who once held beauty as a creed. These have proven, to their own satisfaction at least, that working in the clangor and dust of a mill, tending incomprehensible, driving machinery, was far preferable to stringing a loom of sticks and cords under the shade of a tree, or stamping patterns by hand on bits of cloth in some variegated bazaar or beside some idle river's flow. The balance of trade in Lancashire, in Holland and England or Massachusetts, was supposed to compensate for the looted lives of the lineal descendants of unnumbered craftsmen.

I admit, with candor and bitterness, that the powers of darkness have done their uttermost to the ends that charm might vanish, but they have not quite succeeded.

There is another, perhaps an even more precious delusion in regard to color. Most people, and here again I include myself, have believed almost as a tenet of faith, that in the so-called natural dyes there was a subtle beauty, impossible to achieve in any synthetic or chemical formula. To get color from roots and leaves, from bark, clays, flowers and fruits, seemed more in accord with nature's scheme of loveliness than to take color from the test tube of the chemist. To treasure this illusion with consistent conviction, it is necessary to avoid even a slight knowledge of the actual practices of these natural dyers. The substances used were varied and often of an unpleasant character. Acquaintance with the methods of application would make the laboratory of the modern dye expert seem a comparatively delightful location. The dyers' quarters,

INDIA

1—Painted cotton hanging, early Seventeenth Century, from ruined city of Amber. *(Page 70)*
Brooklyn Museum.

2—Detail of Mogul painting on cotton, Sixteenth Century, showing use of cotton fabric as canopy from Romance of Amir Hamzah in reign of Akbar the Great. 1556–1605. This forms one of a series of illustrations painted on cotton, nine of which are in the Brooklyn Museum. These paintings are believed to have powerfully influenced William Morris in his decorative design. *(Page 70)*
Brooklyn Museum.

3—Painted cotton hanging from ruined city of Amber with portrait of Sir Thomas Roe, British Ambassador of the Court of Jahangir at Agra, 1615–1618. *(Page 70)*
Brooklyn Museum.

4—Seventeenth Century painted and stamped cotton wall hanging from southern India. *(Page 70)*
Brooklyn Museum.

PLATE 12

1

2

3

4

in ancient Asia and in Medieval Europe, were earnestly avoided by the discriminating and sensitive.

Freely and joyously, I admit that our modern fabrics, at least the vast majority of them, do not compare in beauty with the ancient webs; but I am convinced that this is not because of the chemistry of color, but because the chemist is seldom an artist, and the artist must control the application of dye to fabric, as the chemist must control the formulas which create the possibilities of the artist. When this happy balance exists between functions and individuals, modern dyes are capable of producing chromatic effects that will meet the requirements of the most exacting taste.

This is a kind of confession, for but recently I held another point of view. To me the great dye industry of today was one of the most destructive agencies for loveliness that the ingenuity and deviltry of man had conceived, and I regarded the promiscuous and uncontrolled introduction of aniline dyes into India as one of the seventy deadly sins of our day. Many fabrics coming from India, ghostly reminiscences of former beauty, distorted into garish ugliness, seemed to support my belief. The old craft of the dyers, where beauty had matured for centuries, had been swept aside by these sinful pastes and powders and a devil's holiday made of color.

Recently, as a part of India's rebellion against the West, which included a partial embargo against British manufactured goods, the old craft guilds were in some measure revived. In a country as large as India, there always had remained a few individuals who insisted on the old traditional designs in fabrics, and these were the saving spark.

With humility proper to the occasion, I admit that

the National Aniline & Chemical Company, one of the world's largest dye manufacturers, heaped coals of fire on my head. My criticism of their colors had been active and sustained, and I had often shown different members if this organization the ancient colors, in proof of my criticism. They sent to me from Bombay a collection of tied and dyed silk and cotton saris that are utterly worthy of the best traditions of the ancient East, both in craftsmanship and in color.

It is, therefore, with a deep satisfaction, that I conclude this brief chapter on the cotton arts in India with this reflection. The world may again have every beauty that graced the golden centuries, that brought charm and gaiety to the Europe of those periods, if it but possess the will to this beauty. It has not perished, it but awaits the demand that launched ten thousand keels.

INDIA

1—Indian printed cotton hanging of the Eighteenth Century.   (*Page 70*)
Brooklyn Museum.

2—Eighteenth Century painted and printed wall hanging from southern India.   (*Page 71*)
Brooklyn Museum.

3—Detail of painted hanging from Amber, Sixteenth Century or early Seventeenth Century.   (*Page 70*)
Brooklyn Museum.

4—Detail of resist dyeing, the triumph of Bacchus from Egypt about 400 A.D.   The Louvre Museum.   Discovered by M. Gaget at Antinoe.   A similar process is described by Herodotus, 450 B.C.   (*Pages 64, 75, 76*)

PLATE 13

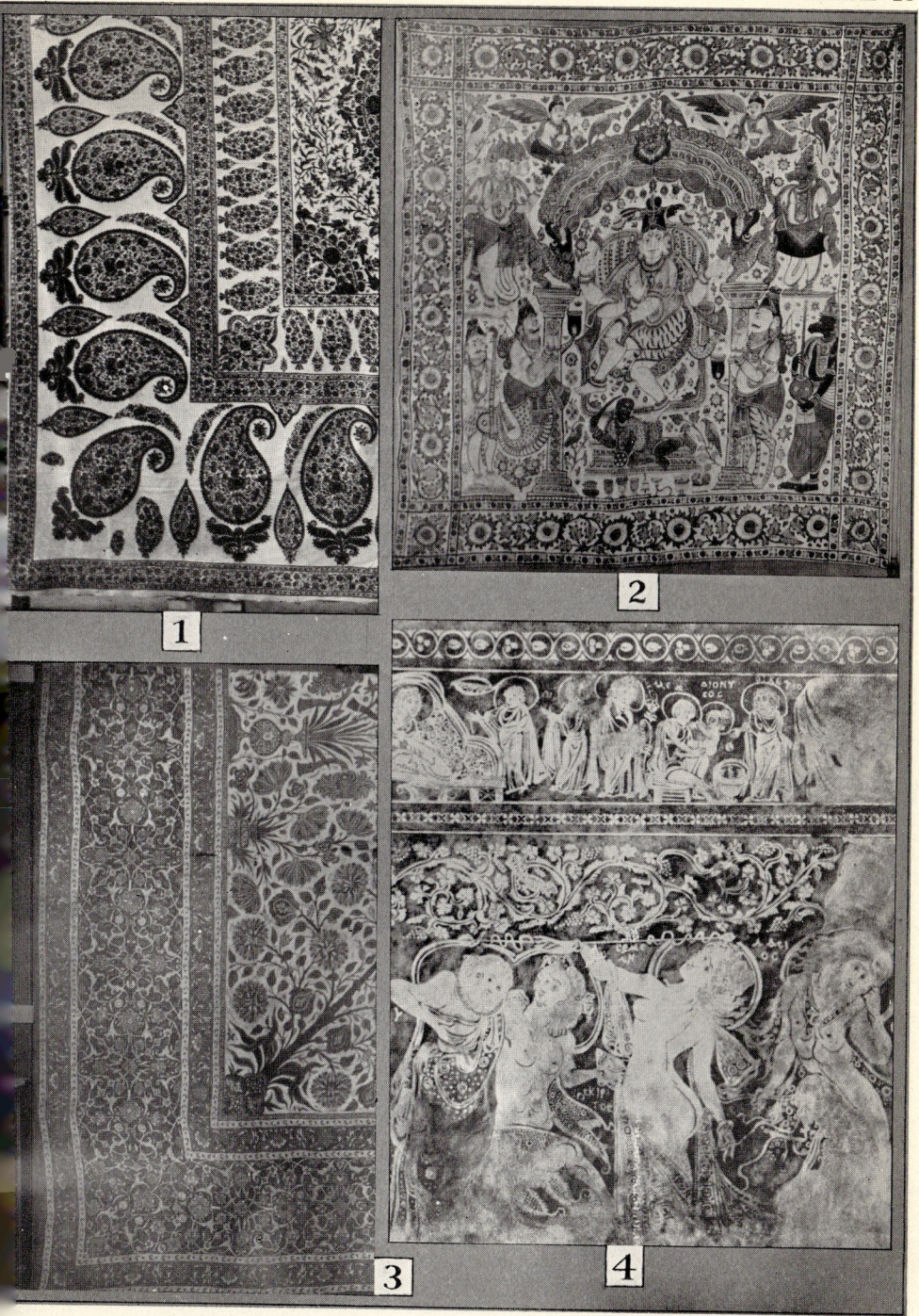

CHAPTER VII

EUROPE

COTTON was cultivated on the Greek mainland in the time of Alexander the Great and in the Saracenic Islands of the Mediterranean from the Eighth Century on. But Spain, under the Moorish Caliphs, was the most important early European center of the fiber. Tradition runs to the effect that one of the cultured Mohammedan rulers presented a cotton embroidered mantle to Charlemagne of France before 814 A.D. It is well established at least, that in the reign of Abdrahaman III, 912–961 A.D., the arts of cotton cultivation and conversion, together with other Oriental crafts, reached a high distinction in southern Spain. In the ancient cities of Cordova, Seville, and Grenada, cotton weaving and dyeing is said to have compared favorably with that of Bagdad and Damascus. The Moors even made paper from cotton long before any other European city understood the art. Today cotton grows wild in the fields of Valencia in memory of more gracious times. Barcelona was famous for her rough fustians and sail cloths, and two streets in this ancient city, "Cotoners Velle" and "Cotoners Nous" derived their name from the days when cotton was an important economic factor.

Moorish Spain had little contact with Christian

Europe because of the almost constant warfare and the bitter religious differences, and when the Mohammedans were expelled their crafts went with them, leaving only a trace behind, although their influence on design was both powerful and persistent.

The Oriental commerce and arts which took root in Italy, in the early centuries of the Middle Ages, came directly from Byzantium, Syria and the Phœnician towns captured by the Crusaders. Before this there was trade in classical times, which was never wholly interrupted. Whether or not cotton appeared in this dim commerce we can not say, but there is no question that the Italian merchant princes were the first to introduce the cotton fiber into Europe generally. Venice is said to have been the first city to have manufactured cotton fabrics, although the earliest record of cotton fiber was in Genoa, where in the year 1140, cotton from Antioch was weighed on the public scales along with the cottons from Alexandria and Sicily. Cotton was grown in early times in Apulia, Crete, Sicily, Cyprus and Armenia, but the fiber was rated as of lower quality than the Levant or the Indian cottons, which came by way of Alexandria.

Cotton fiber was a regular article of commerce between Venice and Ulm as early as 1320, and soon spread to other cities in southern Germany. It was fashioned into cotton and linen and cotton and woolen fabrics, known as "barchents," "fustians," "ripplecht," and "gehorte." Unquestionably in the late Middle Ages, Germany led all Europe in the production of this character of merchandise. England, in the Fourteenth and Fifteenth Centuries, imported large quantities of these fabrics, and there was as well an extensive trade between Antwerp and Venice in printed cottons, both of Indian and European manufacture. England alone bought from the cities of Germany in the Fifteenth Century, six hundred thousand crowns' worth annually of barchents and fustians.

Block printing and perhaps resist dyeing in a single color were known in Europe before the introduction of Indian calico. A few specimens are preserved in European museums of blue and white prints on cotton and linen fabrics of earlier dates. In the Fourteenth Century, Italy expelled the Anabaptists, first of the great Protestant sects, and these immigrants helped to build up in Central Europe that body of arts we call today by the misleading term of peasant arts. We know these people were skilled printers of cotton and linen fabrics and to have imitated in prints the blue and white embroideries of Italy. I am inclined to believe that the blue and white prints were a reminiscence of earlier eastern contact through classical Italy. It seems difficult to associate this form of expression with the multicolored calicoes of the Seventeenth Century.

The earliest forms of European carved wooden blocks for printing on fabric we possess are, however, obviously copies in form and design of the Oriental prototypes of the Sixteenth Century. They are irregular in size and shape and each one is a complete unit. Later the wood engravers and etchers of Europe influenced the printers of fabrics and we find rectangular stamps carved in complete and rigid patterns, composed of many units. The last phase, and the one still in use, consists of a number of blocks, each carved in the details of a single color and all so carefully related, as by successive printings to produce many colored designs.

The Anabaptists were the ancestors of our Mennonite Quakers who brought to Pennsylvania a knowledge of glass making, glazed pottery, weaving of double cloth coverlets and the printing from wooden blocks on fabrics, and we may still discern in these modest Colonial arts a reflective loveliness of the cultured races of antiquity.

A list of the cloth printing establishments in Europe and England will give some idea of the importance of the cotton trade in the Seventeenth and Eighteenth Centuries. Plants were founded in Augsburg in 1688, in Richmond near London in 1690, in Neuchatel in 1716, in Schwechat near Vienna in 1726, in Glasgow in 1732, in Hamburg in 1737, in Zschopau in 1740, also in Schlesia and Rehin-Prussia, in Berlin in 1741, in Mulhausen in 1746, in Plauen in Vogtlande in 1750, in Wesserling in Els in 1760, in Grossenhai in Sachsen in 1763, in Preston near Manchester in 1764, also in Lancashire, in Heidenheim a. d. Brens in 1766, in Chemnitz in 1770, in Jouy in 1776, in Kosmanos in Bohemia in 1778 and in Iwanowo in Russia in 1780. These plants were obviously the result of the trade with India.

Such skill in fabric printing as existed in Europe before the introduction of Indian chintzes and calicoes was simple in pattern and in a single color. Our ancestors used woad, a plant containing a weak dye not unlike indigo in chemical reaction, and in earlier times perhaps the juice of the Mediterranean shell fish that produced the royal purple of Tyre. Indigo was discovered by Marco Polo in 1300 and was probably introduced into Europe a little later in small quantities. In the Seventeenth Century there are records of very large shipments which indicate that the cotton and linen industry in dyeing and printing had become very extensive in Europe. Indian madder was used from

COTTONS OF EUROPE

1—Spanish embroidery of the Sixteenth Century, on two thicknesses of cotton, embroidered in yellow silk floss. Figures are both European and Indian. (*Page 82*)
Brooklyn Museum.

2—Italian block printed cotton of the Nineteenth Century. (*Page 87*)
Metropolitan Museum of Art.

3—Italian block printed cotton of the Nineteenth Century. (*Page 87*)
Metropolitan Museum of Art.

4—German printed cotton of the Eighteenth Century. (*Page 84*)
Metropolitan Museum of Art.

5—German print, Seventeenth Century, Oriental in design. (*Page 83*)
Metropolitan Museum of Art.

6—Modern and Eighteenth Century single color resist block printing from Kremnitz, Hungary. (*Page 84*)
Brooklyn Museum.

7—Printing blocks from Germany and southern Hungary, about one hundred years old. (*Page 83*)
Brooklyn Museum.

PLATE 14

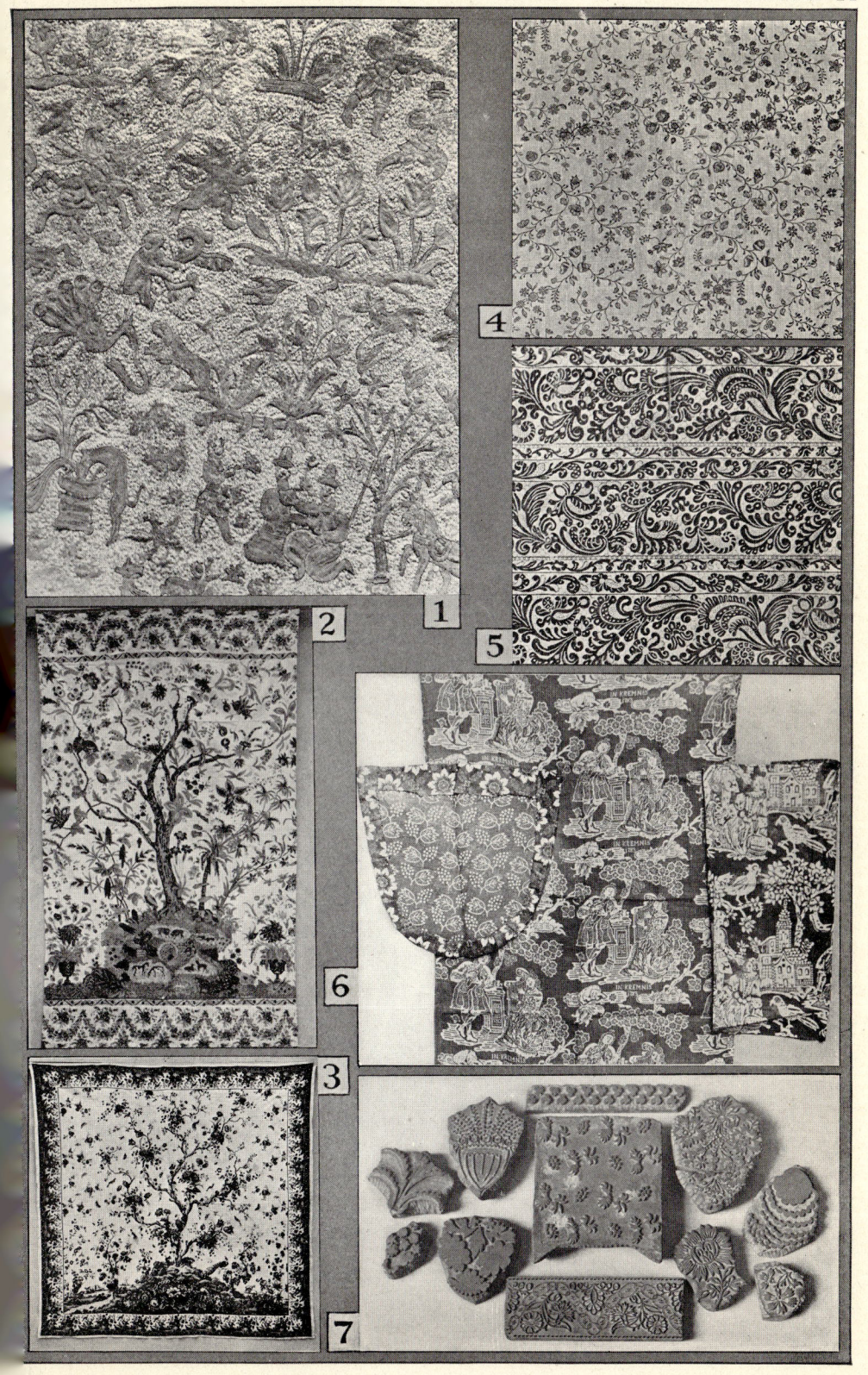

classical times to produce red and after the conquest of Mexico, the cochineal insect was imported and later raised in Africa and the Near East.

After the voyage of da Gama to Calcutta in 1497, Portugal built up at once a large and profitable trade with the Indies, including cotton and cotton fabrics. The Dutch traded with Lisbon and in this way Antwerp, Bruges and Haarlem became the most important cotton ports of northern Europe, and the old industries revived, indeed they were vastly stimulated by the rich colored chintzes and calicoes which formed a part of each shipment from the East. We do not know much about the manufacture of cotton in Portugal itself, but it is beyond question that the art of block printing grew up among these people and even affected in some degree the arts of India. Spain, after the conquest of Mexico and Peru, found in her vast possessions of the West, sufficient scope for her restless energy. So for almost a century, the descendants of Henry the Navigator monopolized the cotton trade in the Far East.

A few Dutch adventurers and Dutch captains of Spanish vessels, began to voyage among the islands of the eastern seas, carrying on a more or less illicit traffic, bordering often on open piracy. Spain resented this bitterly because of her religious and political differences with Holland, and finally forbade Dutch trade with Lisbon. But the seven seas were then as now a rather wide area over which to enforce authority, and the men, who had met with stubborn courage the armed might of Spain, were difficult to impress with royal edicts. They combined to defend themselves from Spanish vessels of war. Then Spain, mad with ambition, committed national suicide, and the British guns, before which the mighty Armada crumbled into helpless, pitiful wrecks, opened the ports of the East to the merchant adventurers of two rising maritime people.

The Dutch merchant adventurers eventually amalgamated in 1602 into one company, under the title of the Dutch East India Company. In 1587, Drake seized the *St. Phillip*, a Portuguese carack from the East Indies, and English privateers in 1592 captured the *Madre de Dios*, and in her cargo discovered calicoes, lawns, quilts and carpets and other rich commodities, and no doubt bills of lading of such value that English adventurers realized the immense wealth to be obtained through direct contact with India. So a memorial was presented to Queen Elizabeth in 1599, and in 1600 a charter granted to the first British East India Company. This does not mean, that the English preceded the Dutch as traders in the East Indies. There is no doubt that the Dutch mariners were the first in these seas after the Portuguese.

Cotton is mentioned in France in the Fifteenth Century as padding for armor and for some vague millinery purposes, also for hair nets, and was perhaps used in the rough cloths, mixed with linen, I have mentioned, as being made in southern Germany. The French East India Company was formed in 1664, more than half a century later than the Dutch and English.

The introduction of painted calicoes and chintzes of the East met in French markets, in the Seventeenth Century, a vigorous resistance from the manufacturers of silk and wool. Stringent laws were passed and in a measure enforced, and yet it is apparent that in spite of these prohibitions, and maybe because of them, there was a dangerous demand for the forbidden wares. In Molière's *Le Bourgeois Gentilhomme*, the tailor, attempting to ape the habits of the gentleman, remarks that fashionable people wear chintz dressing gowns in the morning. Early in the Eighteenth Century it was estimated that twenty million francs were spent annually on Oriental calicoes and German and perhaps Italian imitations. At this time there were no printers in France able to make fast calicoes, although there was a forbidden traffic in fugitive pigment colors.

Prussia had similar prohibitive laws in effect until a calico plant was erected in Berlin in 1741.

In 1758, Christoph Phillip Oberkampf, son of a dyer in Weisenband, settled in Paris and entered into partnership with a M. Cottin, who up to that time had never been successful in printing fabrics in fast colors.

Oberkampf began at once the printing of calicoes in fast tints and so successful was he and other printers, that in 1759 all restrictions in regard to the domestic trade were removed. In 1776, Oberkampf formed the new partnership of Sarrasin-Demaraise et Oberkampf and established at Jouy, a little town near Versailles, his famous printing works.

Very rapidly the beauty and excellence of these fabrics attracted buyers from all over Europe, including England. Oberkampf was a great favorite of Marie Antoinette, and in 1787 the unhappy Louis XVI conferred upon him letters of nobility. By shrewd direction and a very skillful manipulation of the laws of exchange in assignats, he kept his organization intact during the period of the Revolution and survived the period of reconstruction successfully. In 1806, Napoleon and the Empress Josephine visited his works and Napoleon decorated him with his own Cross of the Legion of Honor. In 1809, Oberkampf received first prize for his contributions to science and art in the following citation:

"M. Oberkampf began his establishment fifty years ago and naturalized in France the art of painted cotton, which had been built up in Europe from very modest beginnings. M. Oberkampf elevated his manufacture to a great degree of prosperity. He brought perfection in all phases of the industry, be it by the application of chemistry or mechanical processes. Among the new processes should be recognized the engraving of cylinders and plates of copper, the impression of a solid green in a single application, and the use of steam in the process of dyeing."

Once again in 1810, the great Corsican visited this father of the French textile printing arts and planned with him a joint war against the English.

"We will make together," said the Conqueror of Austerlitz, "a rude war against the English, you by your industries and I by my armies."

Most of these famous prints are in single colors, stamped from engraved copper plates. Later the cylinder printing of England was introduced but did not fit the genius of the French workmen so perfectly as the more artistic method.

These brief comments are little more than a hasty sketch of the history of cotton in continental Europe. I have done scant justice to the arts of the greatest industrial and cultural significance. There is no doubt that in the Eighteenth Century, and perhaps even before, the vigorous realistic expression of Europe, born of the Renaissance, strongly affected the styles of Indian printing. There is no question that France achieved a higher distinction in her own inimitable fabrics than I have suggested. The supremacy of French handcraft looms and in printed cottons of distinction today implies a greater share in the artistic history of the fiber. If we owe much to India (and this debt is beyond question) we owe an almost equal debt to the French craftsmen, who have kept alive the beauty of texture, design and color in cotton. France unquestionably borrowed technology and chemistry from England and Germany, but this debt has been more than repaid by the sustained artistry and traditional good taste of her craft designers.

Other factors, however, make it seem wiser to continue the story of calicoes and cottons in the history of that nation most concerned in the later commerce and technology of the fiber.

In the great Elizabethan age, the vigorous mariners of England began that series of audacious voyages and expeditions, which, in a short century, placed the commerce of the world in their control. French, Portuguese and Dutch adventurers still played important parts in their Indian dominions, but to England fell the chief rôle, so the next century and a half of cotton history becomes merely a detail in the growth of the British Empire.

CHAPTER VIII

ENGLAND

THE pre-eminence of Great Britain in the modern cotton industry, her splendid mechanical contributions in the Eighteenth Century, the vast acres of cotton plantations that look to her as a market for their product, make it difficult to imagine a time when cotton was unknown in England, or even of little importance. It is not, however, until the latter part of the Thirteenth and the forepart of the Fourteenth Century that we have any record of cotton in England at all. The oldest record of cotton in the British Islands is contained in the Compotus or inventory of Bolton Abbey. Cotton is here referred to as being used for candle wicks. In a poem, the "Siege of Caerlaverock," dating from 1300, a passage runs: "Maint riche gamboison garni de soie et coton," ("Many a rich doublet trimmed with silk and cotton"). The "Compotus Earl of Derby," dating from 1381–82, speaks of cotton thread and of six pounds of cotton wool.

These scattered accounts do not determine perhaps the first actual appearance of the fiber in England. Cotton may have been an unconsidered article of trade with Italy and even with Moorish Spain and passed unnoticed in these non-statistical ages. Within the first generation of this century, the cotton and linen fustians and barchents of southern Germany became very important articles of trade with England, and attracted the attention of the poet Chaucer and others; and there seems to have been a strong disposition to have such fabrics manufactured in England.

At about this period or a little earlier, Flemish weavers were introduced into England through the patronage of the Crown. England was at this time almost entirely an agricultural nation, her chief exports being wool, which was shipped to the busy looms of the low countries.

It is a great testimonial to the splendid stewardship of the British kings that they sought to build up in their realms the textile arts, and to make little England as independent in manufacturing as she was in political life from her wealthier and more powerful neighbors on the continent. From the time of Edward IV to the end of the Elizabethan period, textile workers were openly and secretly encouraged to settle in England and given certain privileges and a sure protection. Yet of more importance than these worthy efforts were the fierce religious wars in Holland and Flanders, and the crowning horror of St. Bartholomew's Eve in the tragic streets of old Paris in 1572. The Revocation of the Edict of Nantes was the last episode responsible for hundreds of thousands of refugees from the continent, mostly skilled artisans, seeking refuge in England in a little over a century. Many, perhaps the majority of these, were textile workers.

In consequence, at the dawn of her rise as a great naval power, she was well advanced in the textile arts. In the late Sixteenth Century she was making the cotton and linen fustians formerly imported from Germany in her own midland countries and even exporting her surplus.

England's modern cotton history begins while the memory of the terrible Armada of Spain was still fresh in the public mind, and is closely connected with the capture of Spanish and Portuguese vessels from the Far East, mentioned in the previous chapter.

The founding of the British East India Company, in 1599, was followed by a vigorous generation of Oriental trading, which changed not only the industrial, but the artistic and social life in England. So important did this cotton trade from India and the cotton and linen mixtures of Lancashire become, that the growers of wool and the manufacturers of woolen fabrics made vigorous protest to Parliament for redress. In 1621, twenty years after the founding of the British East India Company, the wool merchants made the following complaint:

"For about twenty years divers people in this kingdom, but chiefly in the county of Lancaster, have found out the trade of making fustians out of a kind of down, being a fruit of the earth growing on little bushes, or shrubs brought into this kingdom by the Turkey merchants from Smyrna, Cyprus, etc., but commonly called cotton wool and also of linen yarn, and not part of the same fustians of any wool at all. There is at least 40,000 pieces of fustian of this kind yearly made in England, and thousands of people set on working of these fustians."

Anyone familiar with the history of England at this period, may well believe that such petitions were taken with great seriousness. Wool and all that concerned wool was of vital moment to England. The modern political philosophy of Adam Smith, with free trade and the recognition of labor as a commodity, had yet to be born. We may, therefore, assume that the introduction of cotton merchandise into England was looked upon from the start with scant favor by those not engaged in the trade.

There was, however, a growing interest in outland cotton trade, as a note from John Gourney at Patania in 1614 indicates. He describes certain difficulties encountered in securing cotton goods by the British East India Company:

"When this man (the native official) is feed by weavers and such as seek to trade with us with about eight or ten per cent., they may freely come and bring us wares, and, besides what the Governor cometh to knowledge of, must yield at least ten per cent. more; and sometimes men have been taken and accused of having gotten much by trade, and after many blows and a long imprisonment pay a forfeit of all the money they have taken. This makes poor men bring their paintings in huggermugger and in the night, as thieves do their stolen cloaks to brokers."

In 1649 from Ahmedabad a letter from John Tash, indicates certain further difficulties experienced with the local rulers.

"Our chints and tappichindaes, which were the most considerable part of our Bantam investment, were delivered to the workmen so seasonably that they were all returned painted before the rains; so that there was as then nothing wanting unto them but washing, which might have been performed in eight days' time (they being to that purpose returned to them upon cessation of the raines) but our Governor not suffering them to work in the river hath been an exceeding impediment to the business, whilst the chinters have neglected their work to attend upon the Durbbar and sollicite redress, which for consideration of 250 rupees was once granted to them; to the payment whereof they had no sooner consented than our base, unjust and worthless Governor raysed the sum to 1000 rupees with further obstruction."

English and Dutch rivalry was very strong, as might well be expected in this market, and in 1664 the Dutch expelled the British East India Company from Calcutta and in 1682, on the fall of Bantam, the English withdrew from Java. An extract from the Dutch records is as follows:

"As to our relations with the English, these are rather poor, for we can not come together without the English shaking their tail. Your Excellency may consider how they burst for spite seeing our success in trade.

"The English have beheaded their King and are intent upon breaking with all their neighbors especially with us, in order to secure supremacy of the sea and the monopoly of the trade. This cannot be allowed by the Dutch Nation.

"We seem to be at war again with the English."

It was apparently the custom for the European traders in India to buy the cloths from the weavers and to send these out to the dyers, printers and finishers. This practise was attended with great difficulties, due to the natural indolence of white men in a tropical country, lack of knowledge of native customs and the language and the traditional greed and tyranny of the native rulers.

From time to time the terrible famines and plagues of India interrupted the trade in frightful earnest. A letter from Fort St. George on the Coromandel Coast

ENGLAND

1—Poster in the exposition of "Ancient and Modern Cottons," organized by the author for the National Association of Cotton Manufacturers in 1923. *(Page 205)*
Christine Chaplin.

2—Blouse, Seventeenth Century, embroidered in design suggested by Indian cottons. *(Page 92)*
Victoria and Albert Museum, London.

3—English print, Eighteenth Century. *(Page 103)*
Metropolitan Museum of Art.

4—English print, Nineteenth Century. *(Page 103)*
Metropolitan Museum of Art.

5—English print, Seventeenth Century. *(Page 103)*
Metropolitan Museum of Art.

6—Jacobean embroidery suggesting Indian design. *(Page 103)*
Metropolitan Museum of Art.

7—English print, Eighteenth Century. *(Page 103)*
Metropolitan Museum of Art.

8—English print, Eighteenth Century. *(Page 103)*
Metropolitan Museum of Art.

9—Elizabethan embroidery showing the influence of calico design on English craft arts. *(Page 92)*
Metropolitan Museum of Art.

10—Fanciful drawing of Scythian lambs by German artists in Fifteenth Century. *(Pages 5, 63)*

PLATE 15

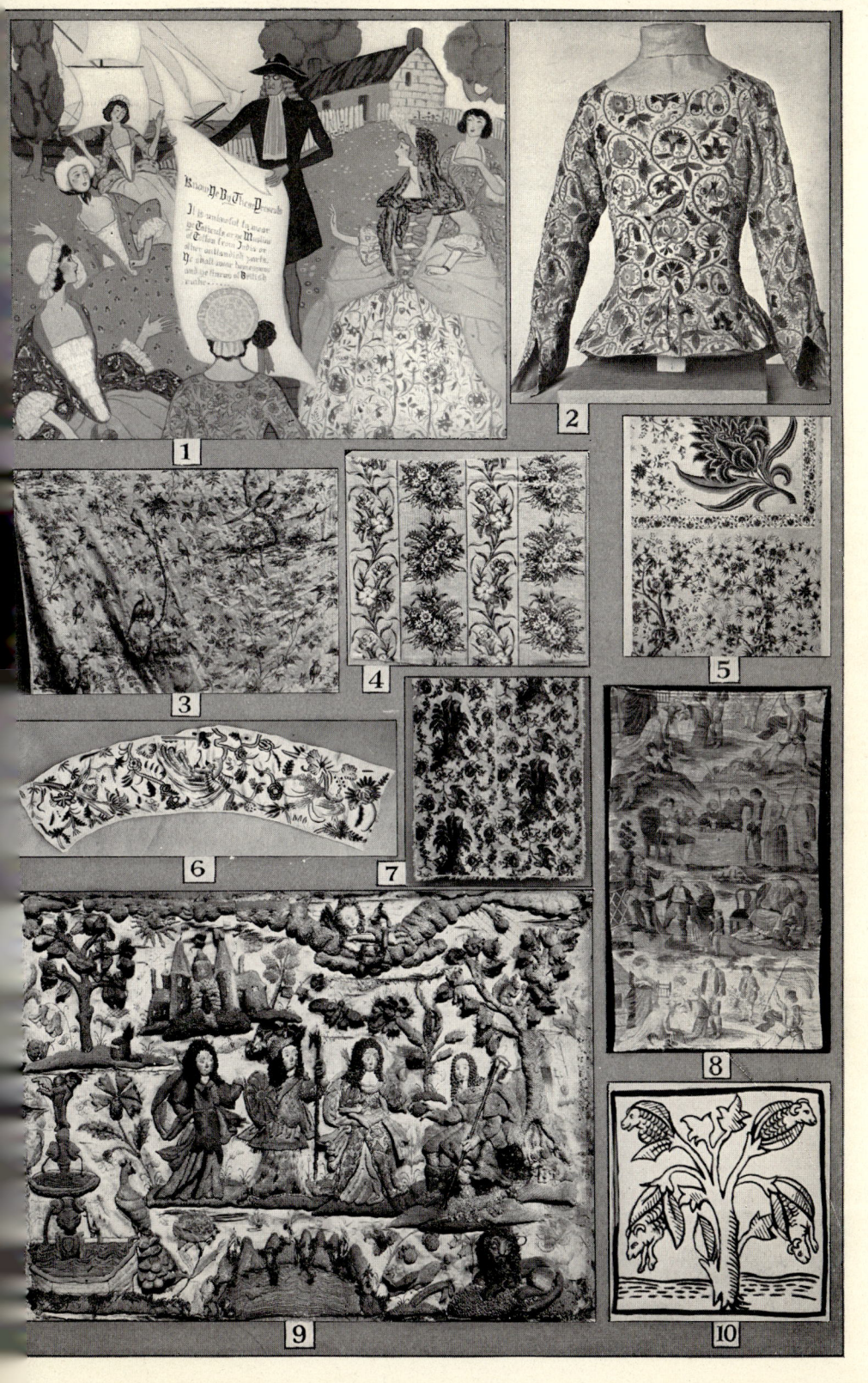

gives us a lurid picture of one of these epidemics in 1687:

"Weavers and washers all dead or gone.......... 35,000 dead at Madraspatam and 6,000 families removed................ Whole tribes of mechanics extinguished with their arts. There is but one dyer surviving in the Bay."

The first British printing plant was organized in Richmond near London in 1690, and while it was a very modest affair, it was rapidly followed by others. The act of William III, in 1696, prohibiting the importation of woolen goods from Ireland into England, but admitting free of duty the linen wraps used by the weavers of England in their cotton and linen cloths gave great encouragement to this domestic trade, although an obvious injustice to Ireland.

The effect of printing plants in England was soon felt in the Indian trade. Evidently there were merchants even in those days who regarded cheapness as the prime quality in merchandise.

A letter to Bombay in 1711 contains the following passage:

"We do find the Bales of Chintz are of the worst cloth and prints that ever came.................... extreamely dear....................... the Printing stands us in as much as the cloth which is a great abuse upon us for our People here will do it at ½ that price and better colours and patterns."

This may, of course, have been what is known as a shrewd business letter. Or, it may be that even as early as this English merchants were forcing Indian craftsmen to carry out English ideas of design with the natural results.

The growth of the trade of Indian printed calicoes

in England is only roughly suggested in the following statistics. These do not include other forms of cotton goods imported, nor the private trade which was, while of a semi-legal character, still very extensive.

In 1671, thirty-four thousand pieces of chintz were imported into England. The pieces were only a few yards in length.

In 1681 English merchants secured designs from Holland to be copied in India. It is rather curious to note even at this time the insistence upon originality in pattern. One letter contains the following passage:

"Every one desiring something that their neighbours have not the like."

By 1683 the demand for calico was beyond all belief. It had become the style and everyone wanted it. The following quotation from a letter to India:

"You cannot imagine what a great number of the Chintzes would sell here, they being the ware of gentlewomen in Holland. Make great provision of them beforehand; 200,000 of all sorts in a year will not be too much for this markett, if the directions be punctually observed in the providing of them................"

Again in 1686 this letter is of interest:

"You may exceed our former orders in Chintz broad of all sorts, whereof some to be of grave and cloth colours, with the greatest variety you can invent, they being become the weare of ladyes of the greatest quality, which they wear on the outside of Gowns Mantuoes which they line with velvet and cloth of gold."

In 1700 the protests of the woolen manufacturers and sheep farmers of England forced Parliament to forbid the selling of cotton goods in England, and in 1712 a further act prohibiting the use of all printed goods, cotton or otherwise, was passed by Parliament. These laws were, however, not as efficacious as desired, for in 1721 a further act was passed imposing a fine of £5 on the wearer and £20 on the vendor of cotton goods. When the difference in money values between these times and the early Eighteenth Century are considered, it must be admitted that these fines were very serious matters.

That they affected the Indian market there can be no doubt. A letter to the Governor and Council at Fort St. George in 1704, four years after the passage of this first law, indicates this situation very clearly:

"We are sorry we can give no great encouragement at present to the fine paintings which we are sensible is brought to great perfection with you. The chief expense (i.e. sale) of that commodity was in England, which now by the Prohibition is taken away, and abroad they turn to no account..............."

To show how vigorous was the protest of the British wool growers and their paid literary defenders, a quotation from a pamphlet "The Ancient Trades Decayed and Repaired Again," in 1678 is pertinent:

"This trade (the woollen) is very much hindered by our own people who do wear many foreign commodities instead of our own: as may be instanced by many particulars, viz. instead of green sey that was wont to be used for children's frocks, is now used and Indian-stained and striped calico; and instead of a perpetuana or shalloon to lyne men's coats with, is used sometimes a glazened calico, which in the whole is not above 12 d. cheaper and abundantly worse. And sometimes is used a Bangale that is brought from India both for lynings to coats and for petticoats too; yet our English ware is better and cheaper than this, only it is thinner for the summer. To remedy this, it would be necessary to lay a very high import upon all such commodities as those are, and that no calicoes or other sort of linen be suffered to be glazened."

A second pamphlet in 1696 *The Naked Truth, in an Essay upon Trade*, runs as follows:

"The commodities that we chiefly receive from the East Indies are calicoes, muslins, Indian silks, paper, saltpetre, indigo, etc. The advantage of these commodities is chiefly in their muslins and Indian silks (a great value in these commodities being comprehended in a small bulk), and these becoming the general wear in England. . . .

"Fashion is truly termed a witch; the dearer and scarcer any commodity, the more the mode. 30 shillings a yard for muslins, and only the shadow of a commodity when procured!"

Even the great Defoe was induced to lend his genius to the vain attempt to stem the tide. He is, however, writing evidently at a time after the passage of the laws had in a measure ameliorated the condition of the home manufacturers.

"The general fansie of the people runs upon East India goods to that degree that the chints and painted calicoes, which before were only made use of for carpets, quilts, etc., and to clothe children, and ordinary people, become now the dress of our ladies; and such is the power of a note as we saw our persons of quality dressed in Indian carpets, which but a few years before, their chambermaids would have thought too ordinary for them; the chints was advanced from lying upon their floors to their backs, from the foot-cloth to the petticoat; and even the Queen herself at this time was pleased to appear in China and Japan, I mean, China silks and Calico. Nor was this all, but it crept into our houses, our closets, and bed-chambers; curtains, cushions, chairs and at last, beds themselves, were nothing but callicoes or Indian stuffs; and in short, almost everything that used to be made of wool or silk, relating either to the dress of women or the furniture of our houses, was supplied by the Indian trade."

"Above half of the (woolen) manufacture was entirely lost, half of the people scattered and ruined, and this by the intercourse of the East Indian trade."

In France equally broad minded legislation had been passed at the instigation of the silk industry. Holland, however, had kept her ports open and did with both France and England a highly profitable business in smuggled goods. The British East India Company as well connived at its own particular brand of smuggling, and the English captains of vessels conducted a very profitable if somewhat risky trade. One author in a burst of prophetic eloquence sums up the situation neatly in the following phrases.

"Two things," says this writer, "among us are too ungovernable, our passions and our fashions.

"Should I ask the ladies whether they would dress by law, or clothe by act of Parliament—they would ask me whether they were to be statute fools, and to be made pageants and pictures of? They say they expect to do what they please—so they will wear what they please and dress how they please."

With such moral support, in spite of all those laudable efforts on the part of British wool manufacturers and Parliament, the cotton trades could not be discouraged, but continued to increase. Dr. Stukely, in 1724, describes Manchester as "the largest and most prosperous village in England. 2400 families were engaged in textiles and there was not only trade with the rest of England, but the beginning of British export trade as well."

Defoe in 1727 mentions cotton manufacturing as the cause of the great prosperity of this town.

Bolton was of somewhat less importance in this trade, but still a central market place where salesmen from Ireland on market days offered the linen warps spun in Belfast. Cotton lint was given to the cottage spinners to be returned as yarn, which the merchants distributed to the weavers and the cloth was brought in in bolts on market days, and sent out to be dyed, bleached, finished and printed by the merchants who sought in Lancashire their supplies of these forbidden fabrics.

George Crompton, the oldest son of the great inventor, born in 1781, describes as a child how he was employed in this manufacture:

"My mother used to bat the cotton on a wire riddle. It was then put into a deep brown mug with a strong ley of soap-suds. My mother then tucked up my petticoats about my waist and put me in the tub to tread upon the cotton at the bottom. When a second riddleful was batted, I was lifted out and it was placed in the mug, and I again trod it down. This process was continued until the mug became so full that I could no longer safely stand in it, when a chair was placed beside it and I held on the back. When the mug was quite full, the soap-suds were poured off and each separate dollop of wool well squeezed to free it from moisture. They were then placed on the broad rack under the beams of the kitchen loft to dry. My mother and my grandmother carded the cotton wool by hand, taking one of the dollops at a time on the simple hand cards."

So prosperous and powerful did the cotton manufacturers at Lancashire become that, in 1736, they had the law of 1721 amended, so as to permit the manufacture of mixed calicoes of cotton and linen in England. They agreed, however, with the wool growers and manufacturers that cottons of India should still be excluded. This exclusion of Indian goods seems to have been in a measure effective, for a few years later the great actor, David Garrick, pleaded in a jocular vein with a friend in the Custom House to permit the painted Indian bed curtains, upon which Mrs. Garrick had set her heart, to escape the clutches of the law.

This statute known as the Manchester Act, had an immediate effect upon the prosperity of this city. By 1750 there were thirty thousand people in the Manchester and Bolton districts exclusively engaged in the manufacture of cotton goods, and by 1766, it was estimated that over six hundred thousand pounds worth of merchandise were manufactured in this region in a single year.

From 1750 on the history of cotton in England is largely the history of the calico trade and of mechanical invention and the application of power to machinery. The calico trade, however, preceded the era of invention and was in fact the chief cause, which called it into existence.

In 1719, in a pamphlet entitled "*The Weavers' True Case*" appears the following comparison in the styles of cloths worn in England before and after the great cotton invasion.

"Let us cast our eyes backward fifteen years (that is to say to 1704, when the prohibition of calicoes decorated in India had been in force four years), and see with what commodities our womenkind were then clothed; we shall see that our women among the Gentry were then clothed with fine English brocades and Venetians, our common Traders' wives with slight silk Damasks, our country Farmers' wives and other good country dames with worsted Damasks, flowered Russels and flowered Callimancoes, the meanest of them with plain worsted stuffs. Whereas now those of the first class are clothed with outlaw'd Indian Chints, those of the second with English and Dutch printed Callicoes, those of the third with ordinary Callicoes and printed Linnen, and those of the last with ordinary printed Linnen."

Contrast this passage with one which appears in the *Gentlemen's Magazine*, on March 14th, 1754, eighteen years after the passage of the Act, which permitted the printing of calicoes in Great Britain.

"Mr. Sedgwick, a very considerable wholesale trader in printed goods, had the honour to present her royal highness the Princess of Wales with a piece of English chints of excellent workmanship printed on a British cotton, which being of our own manufacture, her royal highness was pleased to say she was very glad we had arrived at so great a perfection in the art of printing, and that in her opinion it was preferable to any Indian chints whatsoever, and would give orders to have it made up into a garment for her highness' own wear . . . as an encouragement to the labour and ingenuity of this country."

The great demand for all kinds of printed fabrics aroused the mechanical genius of England to discover more rapid methods of applying patterns. At first carved wooden blocks were employed. This method has never been excelled in artistic value and is still in use. This was followed by more intricately etched copper plates, used in a press not unlike that used to print books. Both of these methods were intermittent and required too great a degree of hand labor. Finally the idea of engraved copper cylinders was thought of, suggested perhaps by the rollers used in making cotton rovings.

This invention had an immense and immediate effect on the entire calico business in England and increased production enormously. It is believed that Charles Taylor and Thomas Walker printed from wooden cylinders before; but in 1770 Thomas Bell, a Scotchman, used the first engraved copper cylinders for this purpose. He is said to have sold his machine to the firm of Livsay, Hargreaves, Hall & Co. about 1785, but this must have been an improved version of his older machine. The description of Bell's machine is as follows:

"A polished cylinder several feet in length (according to the width of the piece to be printed) and three or four inches in diameter, is engraved with a pattern round the whole of its circumference and from end to end.

"It is then placed horizontally in a press, and as it revolves the lower part of the circumference passes through the colouring matter, which is then removed from the whole circumference of the cylinder, except the engraved pattern, by an elastic steel blade placed in contact with the cylinder. (This was the forerunner of our modern printing machines.)

"A piece of cloth may then be printed and dried in one or two minutes which, by the old method, would have required the application of the block 448 times."

In O'Brien's famous treatise on calico printing, written in 1789, he sagaciously comments upon the fact that the use of roller printing would tend to cheapen the once so desirable product.

"What person would willingly give five or six shillings a yard if their very servants could have an imitation of or what has nearly the effect for two or three?"

During this period, when the calicoes of India and the imitations in England were attracting the attention of merchants, craftsmen and manufacturers, a small group of men were at work upon machines which in the end changed the entire complexion of the cotton industry. I have reserved the consideration of this period for a separate chapter.

CHAPTER IX

EIGHTEENTH CENTURY: THE AGE OF THE MACHINE

THERE is some evidence that the first experiments in textile machinery and the application of power to machinery, were made in the Seventeenth Century in continental Europe. It is doubtful, however, if the British inventors of the Eighteenth Century had any direct knowledge of these first hesitating essays; nor do they in any sense diminish the honor of the actual accomplishment. To all intents and purposes, the great industrial revolution begins in the midland counties of England in the fore part of the Eighteenth Century and ends with the dawn of the Nineteenth. There is no question that the entire movement centers about cotton. With two exceptions every inventor was an Englishman and the great majority were drawn from the ranks of the humblest textile workers.

In this period, practically every textile machine we use today was created. All subsequent efforts have been directed towards improving, enlarging and coordinating these basic ideas. In this century we see as well the first complete demonstration of the idea of the division of labor and serial production.

One continental invention was, however, of such importance as to deserve special mention. The flyer, added to the Indian spinning wheel, which appears on the Saxony wheel, and enters so largely into most types of modern mechanical spinning machinery, was the invention of Leonardo da Vinci, either in the latter portion of the Fifteenth or the first part of the Sixteenth Century. It was probably invented to assist the Italian craftsmen and their followers in southern Germany, to more successfully spin the cotton lint, which was then a regular article of trade between the Levant, the Italian cities and the towns of Flanders.

In the *Journal des Savants* 1678, is an account of a machine to weave by power, invented by a M. de Gennes, a French naval officer, which does not appear to have attracted any attention at that time. There was as well a French device to supply power to two hundred hand spinning wheels at a little later date, but no proof of its application in practical form.

The great point about the English inventors was that they were practical artisans, devoting their attention to the actual production of definite and much needed commodities. They were faced with the necessity of carrying theory into practice and testing ideas against facts.

It is possible that the vigorous commerce of England in the Sixteenth and Seventeenth Centuries may have more ruthlessly destroyed the weaving guilds than on the continent, and thus opened up the way for innovation. This is indicated by the greater specialization in textile production and the inclusion of workers in certain phases outside guild control. But of more importance was the character of the British textile population, as compared to those of continental Europe. There was in the midlands that mingling of closely related peoples of slightly different points of view and traditions, which so usually destroys prejudices and stimulates creative thought.

The history of textile devices before the machine age is not lacking in interest. Great changes had been made in the simple types borrowed from the Orient, and all of these changes had been of a mechanical nature, as aids to the production in quantity rather than in fineness of quality. Nor can we escape the thought that both the machine and the idea of power application are closely related to the mathematical philosophy and speculation of the age. Great inventions in all times are the fulfillment of processes of thought, not the starting point of thought.

There is little direct evidence to prove the point, but a strong assumption exists that both the two barred loom and the spinning wheel were borrowed from the East. There is unquestioned proof that all through northern Europe the warp weighted loom prevailed and existed in some remote regions, even down to our own times. In the fragmentary records of early Europe, no mention of the spinning wheel occurs, but the primitive distaff and whorl method is occasionally referred to. Just when the loom was introduced it is difficult to say. The Fourteenth Century silk loom of England, surely, looks enough like the Indian loom to have been a recent introduction, but it may have come into Europe at a much earlier period, possibly even during Roman times.

As a general rule all European textile implements were heavier than their eastern originals and at a very early date we find them growing more complicated in mechanical detail. This is an interesting indication of that first dawning of technical consciousness, which in no small degree accounts for our material ascendency over all Oriental cultures, and while it emphasizes our craft inferiority, perhaps it illustrates at the same time a higher receptivity for mechanical reasoning.

The first concrete expression of the machine age was the fly-shuttle of John Kay of Bury, son of a small woolen manufacturer of Colchester. I have included a detailed draft of the device in another portion of the narrative, and will, therefore, only briefly describe its effects upon weaving, rather than enter into a too technical description of its parts.

Before Kay's invention, the shuttle (the wooden container of the bobbin of weft) was passed through the opening of the warps from one hand to the other, while the pressure of the feet opened the alternating sheds. (See sketch of loom). Consequently the width of cloth was limited to the reasonable spread between the outstretched hands of the weaver. Since the weaver had to operate the battern or beating up frame of wires as well after each shot of weft, it will be seen that weaving was a rather slow process. Kay wished to make possible the weaving of the broad fabrics of India and to conserve the labor time of his weavers. In perfecting this device, he changed the shape of the shuttle and made certain practical additions to the loom, which prove him to have been not only a man of imagination but of unusual mechanical skill.

His perfected invention was greeted with angry protests from both weavers and masters alike, and he was driven from his house and eventually settled in Leeds. Here his fly-shuttle was shamelessly stolen by the manufacturers of woolens and all compensations denied him. Shuttle clubs were formed, each member pledging himself to help defend any member sued by Kay. In the end, though never losing a suit, he was impoverished and discouraged by his many appearances in court, returned to Bury to work as a millwright and machinist. Here his inventive genius led him to perfect a spinning machine. It is believed that in many ways his invention anticipated those of Arkwright and Crompton. Again misfortune pursued him and he fell a victim to that mad fear of all mechanical devices, which pervaded textile England at this time. Kay's spinning machine was destroyed and his life saved only through a hasty and secret retreat. He sought refuge in France, where he died in poverty in 1764.

It has been estimated, that Kay's fly-shuttle increased the average production of a loom four-fold. Even before this time, there had been a great shortage of cotton weft yarn. Spinning was largely an intermittent occupation of women and children in the small agricultural hamlets, while weaving was becoming more and more the steady occupation of a masculine artisan class. It took about three spinners to supply enough yarn for a single weaver and when loom production was multiplied by four, there arose a dangerous shortage. It must be remembered that at this time there was a great activity in the production of British calicoes and an increasing demand for the rough cotton and linen fabrics, all of which made Kay's improved loom of the greatest economic importance. It could not, however, reach its fullest value until adequate supplies of weft yarn were obtainable.

Attention, therefore, turned to some device to increase the productivity of the spinning wheel. There is even a record of a small reward being offered for such a machine. It will be seen that such a machine, or rather the idea of such a machine, could not have been any too pleasant to hand spinners, who were receiving comparatively high prices for yarn. Nor can it be said that the English worker of that day, had any aversion to resorting to the cowardly method of mob law to destroy his fancied enemies. An inventor was regarded in about the same way as an abolitionist was a half century or so later in the South.

The first mechanical spinning device to make more than a single yarn was the roller frame of John Wyatt. This consisted of two sets of rollers, operating at different rates of speed, which partially formed the roving, the preliminary process in yarn making. Associated with Wyatt was Louis Paul, a German, who had invented a rude carding device, consisting of a semi-circular receptacle, studded with wire points, in which revolved a wooden cylinder covered with wire nails. This machine separated the fibers and laid them parallel. Neither of these machines had any immediate value, although both contained the sound principles which were later perfected by other inventors and incorporated in their machines.

The first wholly practical spinning machine was the Jenny of James Hargreaves, a modest cottage weaver of Stanhill. He had such great difficulty in obtaining sufficient weft yarn of cotton for his busy loom, in competition with the merchant manufacturers of Lancashire, that he made experiments with spinning wheels to increase his supply.

His device gives evidence of a fine practical imagination and sound understanding of the principles of spinning. It consisted of a stout frame of wood supporting two racks of spindles, one containing the partially spun roving, still made by hand or on some modification of John Wyatt's rollers, and the other to receive the finished yarn. From the roving spindles at the rear of the frame, a carriage drew out the yarn, while the spindles revolved. Both passage of carriage and the drive of the spindles were controlled by a wheel with a belt or cord attached to pulleys and gears. When the yarn had been spun, the carriage traveled forward and the yarn spindles wound it up.

This machine is easier to describe than to operate and must have required great skill on the part of the worker. It could only spin a rather coarse, weak weft yarn. None the less, since it contained at first eight and later sixteen spindles, it vastly increased the output of yarn and aroused the jealous wrath of the hand spinners. In the year 1767, a mob broke into his home, wrecked his machine and drove him from town in fear of his life. He settled in Nottingham, where a lack of interest in textiles made him safe from molestation.

In order to earn enough money to begin life anew, he secretly made a few machines for the more powerful merchant-manufacturers of his vicinity and this fact made it impossible for him to maintain his patent in a law suit, a few years later. It is a satisfaction to write that in spite of this, he was reasonably successful in business and when he died in 1778, left his family in easy circumstances.

We now come to the man, who though in one sense not a great inventor, still did most to establish the cotton and in fact all other textile industries on the modern basis.

The genius of Richard Arkwright was wholly practical. In organization as in adaptation he was far in advance of his times, and he stands happily in no need of pity, since nature and training had endowed him with a shrewd common sense and energy adequate to his needs. He is properly regarded as the father of the cotton mills of today, the prototype of the modern mill treasurer.

He was born December 23, 1732, in the beautiful city of Preston, began life as a barber and later became an itinerant buyer of hair for wig makers. Of a clear investigative mind, he absorbed all the wild talk he heard, going from one weaving town to another, in regard to the great and mysterious machines. Unlike his acquaintances, he did not regard these rumors as evil or visionary, but strove to marshal them to the advantage of that excellent young hair merchant, Richard Arkwright. He was familiar with Wyatt's spinning rollers, he probably saw every day of his life the spinning jenny of Hargreaves, and many other partially successful devices and experiments. In his fertile mind, all of these facts, fancies and conjectures became ideas.

Incapable of actually working out his ideas, he guardedly sought the aid of expert clock makers, giving each a separate part to make and assembling these unassisted. It is said that Arkwright received little domestic encouragement. His wife, a practical woman, fearing to spoil a good barber in making a bad mechanic, is reputed to have destroyed his first models, in an attempt to cure his madness.

He persisted none the less and at last formed a partnership, in 1769, with Jedediah Strutt, the well-known inventor of the stocking frame. With the support and aid of Strutt, he so far perfected his device as to patent it. He then proceeded to build and organize his first mill in Cromford, which was completed in 1771.

Here the real genius of the man shows clearly. Each process of manufacture underwent his careful scrutiny. He changed details of practice, coördinated parts to improve the product and increase the yield. He discovered or rather proved how the skill and productivity of each workman and each group increased with specialization. More than this he carefully correlated each machine with the functions of the preceding and following machines. In other words he grouped in one unit all the most advanced devices for spinning cotton yarn and welded them into a factory. The sum total of his efforts was a spinning mill which could easily and constantly undersell all competitors. His famous spinning frame, driven by water power, was the first to make cotton warp as well as weft, and thus to his genius we owe the first all-cotton fabrics made entirely from European spun yarn. So successful was he, that a second mill was built, the Masson, on the Derwent in 1775, a year before the Revolution.

Arkwright and his partners were too powerful to be intimidated by mobs. Consequently rival manufacturers simply hired away his trained assistants, and as far as possible imitated his machines, including the use of water as a motive power. He endured this until in 1781, when he at last resorted to legal action. The case was bitterly contested by both parties. It was proved, however, that Arkwright's machines were but combinations of other devices, the theft of which had been sanctified by time, and he lost his case in 1785, under rather peculiar circumstances.

It was not possible, however, to steal his natural ability and energy. He was so much better both as a manufacturer and a business man than any of his rivals, that when he died in 1792, two years after the first successful American mill was built by one of his former workmen, he was a millionaire.

Of all the famous names in this brief epoch of invention, none has a more romantic or indeed pathetic significance than that of Samuel Crompton. Craftsman and musician, mechanic and dreamer, his invention it was which gave to British manufacturers that final control over the difficult art of fine cotton spinning, which made England the world's dominant figure in this industry. For it was his mule, so called because of its relationship to Arkwright's first horse power frame and Hargreaves' jenny, that made it possible to spin fine yarns, strong enough to weave into the lighter grades of cotton fabrics, suitable for the best calicoes.

Before we briefly consider the character of this remarkable invention, a single incident may go to prove how close we are to this entire age of revolutionary invention.

In Bolton, England, there is today an ancient and gracious manor house, carefully preserved through the generosity of Lord Leverholm as a charming museum. It is known in the quaint Lancastrian dialect as Hall i' th' Wood, or Hall in the Woods. It was once the home of Samuel Crompton and is today a memorial to his genius. Built in three periods, 1480, 1556, and 1668, each wing has been restored and furnished in perfect taste and in accordance with its period. Before these spacious fireplaces, with mechanical clock work to turn the hospitable spits, may well have come some fog bound traveler, with strange tales of the voyage of Portuguese da Gama round the stormy Cape of Good Hope, and his safe return to Lisbon, cargoed with rare essences and cotton of exquisite beauty, or some tale of the bold adventure of our great Genoese, beyond the bleak Atlantic, to that other Indies one day to become so vital in English history. When these events took place, the oldest wing had hardly reached the middle of its second decade.

It must not be supposed that in Crompton's time it was a home of wealth and elegance. The ancient manor had fallen upon evil days and was the dilapidated tenement of the poor. His room, the fireplace he sat before in suspended dreams, even his chair with the broad arm for writing and drawing, have been carefully preserved. Here there hangs a portrait of an oval, sensitive face, a cop of yarn he spun and the memory of that great struggle and triumph which no later tragedy might deny him. This is all. His machine is guarded in the British Museum and his fame lives in British history.

Within a few years of this writing, there were in Bolton, two ancient men, who as children had lived in Hall i' th' Wood. They remembered a shabby gentleman, who enjoyed the unusual privilege of wandering about their garden, for they had been told that he was Mr. Crompton, the inventor. So within the reach of one long life, does this vigorous age touch the man whose genius did so much to build it.

Crompton's invention perfected in 1779, was indeed a combination of the rollers of Wyatt, the jenny of Hargreaves (a small one which he possessed and used, serving as a model) and the practical machine of Arkwright, known only through repute. But the combination thus inspired was a stroke of pure creative genius. Since his time, master mechanics and machine builders have lavished on his invention all of their increasing knowledge of metals and fiber and power application, nor have there been lacking additions of great merit. Yet in principle and indeed in appearance, it is as he first perfected it in his young manhood, nor was there ever any machine which combined the virtue of vast production, while still retaining the delicacy of hand directed operation.

No better description of the early mule is possible than that contained in Murphy's excellent work, *The Textile Industries*.

"Having heard of Arkwright's roller spinning frame, though without knowledge of its structure, Crompton set himself to improving the spinning jenny, introducing the roller spinning principle as an aid in attenuating the rove. On the head of the frame he placed the roving creel, and in front of it two pairs of drawing rollers; he mounted the spindles on a carriage capable of being moved to and fro. As will be evident, this was a complete reversal of the motions of the spinning jenny. John Kennedy, a Manchester manufacturer and friend of Crompton, accurately described the chief merit of the 'mule,' as it came to be named, in a paper delivered by him to the Manchester Philosophical Society in 1830. 'The great and important invention of Crompton was his spindle carriage, and the principle of the thread having no strain upon it until it was completed. The carriage with the spindles could, by the movement of the hand and knee, recede just as the rollers delivered out the elongated thread in a soft state, so that it would allow of a considerable stretch before the thread had to encounter the stress of winding on the spindle. This was the corner-stone of the merits of his invention.'"

There is no record, I have discovered, that tells how fine in count were the yarns spun on Arkwright's water frame. They are simply referred to as coarse and heavy. We do know, however, a few years later in America, on an adaptation of his machine, No. 14 and No. 20 were spun. These yarns correspond to our

EIGHTEENTH CENTURY

1—First European spinning device, showing the flyer subsequently used on spinning wheels and later on machines. Invented by Leonardo da Vinci, 1452–1519. *(Page 106)*

2—First carding machine to lay cotton fiber parallel for spinning. Invented by Louis Paul in Germany, 1738. *(Page 110)*

3—Fly shuttle, a device which increased loom production four times. Invented by John Kay of Bury, 1733. *(Page 108)*

4—Spinning mule, first machine to spin multiple fine cotton warps. Invented by Samuel Crompton in Bolton, England, 1799. *(Page 115)*

5—Plain English loom before addition of fly shuttle. *(Page 108)*

6—Spinning jenny, first machine to spin number of cotton weft yarns. Invented by James Hargreaves about 1764. From model in United States National Museum. *(Pages 110, 134)*

7—Water power spinning frame. First machine to be driven by power and to spin number of coarse cotton warps or wefts. Invented by Richard Arkwright, 1768. From model in United States National Museum. *(Pages 112, 114)*

PLATE 16

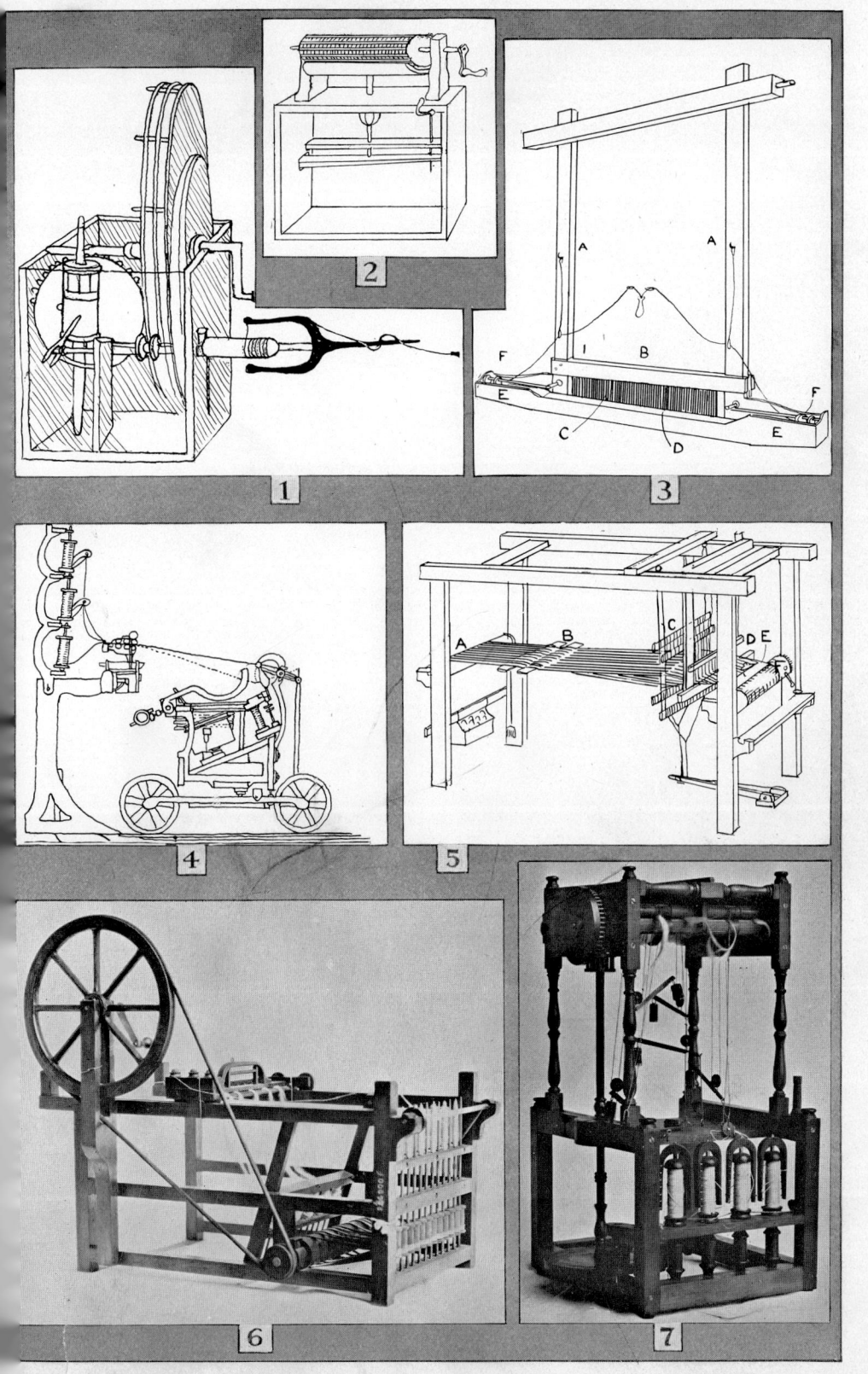

heavy butcher linen or coarse muslins. Crompton himself made good even No. 80's, comparable in weight to the yarns used in fine lawns and nainsook. The hand mule up to our own time was used to make the delicate yarn used in the Calais lace trade and I have modern specimens as fine as 405's or 260,200 yards to the pound or approximately 150 miles. I have also a cop of No. 500 spun for the Jubilee Exposition of Queen Victoria a generation ago. These facts will prove how exquisitely perfect was the original mule for its purpose.

Crompton was as much the victim of his own sensitive nature, as of the sordid greed of his contemporaries. His invention was stolen from him through unredeemed promises of reward and threats against his safety. Long after his invention had brought wealth to many manufacturers he lived in want as a spinner craftsman, attempting to support a large family of young children.

In 1812, a subscription of about £500 was taken up, and a petition presented to Parliament to give him £20,000 in tardy recognition for his great services. But Spencer Percival, the premier of England, on the very morning of presenting the memorial, was assassinated in the House of Parliament, and when the bill was finally placed before the House, it had been reduced to £5,000. This modest sum Crompton lost in a business venture, and lived in his old age on an annuity provided by his later friends. He died in 1827.

As soon as the mule became known, successful efforts were made to increase the efficiency. First one detail and then another was improved. As early as 1790, a Mr. Kelly, strangely located on the River Clyde, applied water power, and by an arrangement of taut and slack pulleys made the movement continuous.

For all its marvelous precision of motion, the mule required highly skilled labor and none knew better their value than the mule spinners. They were the haughty aristocrats of the textile trades, demanding high wages and very special privileges. It is narrated that they wore £5 notes in the bands of their hats and would neither drink nor smoke in the common room of the inns. Worse than this in the eyes of their supposed masters, they refused to train apprentices unless at their own caprice. Even to this day, the strongest craft union in England is that of the mule spinners.

In desperation the baffled manufacturers turned for aid to the great mechanical adapter, Richard Roberts. After five years of experiment, he produced in 1830 the modern self-actor mule, and since this machine could be operated by less skilled mechanics, the arrogance of the early craftsmen was in a measure curbed.

There is but a single name to add to the list of the great inventors, that of Edmund Cartwright, graduate of Oxford, clergyman, writer and inventor of the power loom.

I have already mentioned the drawing of a power loom by de Gennes in France in the Seventeenth Century. The two Barbers in Scotland in the latter half of the Eighteenth Century also made unsuccessful experiments. It is doubtful, however, if he knew of either of his predecessors. He was a man of independent means, with a gift for mechanics, which in a poorer man had surely reaped a greater reward. He made his first power loom in 1785, and later made a second improved model. Neither was wholly successful, the movement that shot the shuttle between sheds, being dangerous, violent, often destroying the warps and occasionally blinding the weaver.

Cartwright at last tired of his experiment with looms, turned his attention to a wool comber, then to a dough mixer, the principle of which is still in force in bakeries, and finally to agricultural implements far in advance of his age.

He seems to have been an amiable, busy man, delighting in mechanical experiment, but not particularly gifted with the persistence essential to success. In 1808, he petitioned Parliament to grant him the £30,000 he had spent on the power loom and actually received £10,000 in recognition of his services.

Dr. Jeffrey of Paisley, Scotland, and Andrew Kinloch of Glasgow, also made experiments in the same direction. Manufacturers very soon adopted this machine, and in 1812 the machine breaking riots were especially directed against it. It remained, however, for W. H. Horrocks and the famous Richard Roberts to bring this machine to something nearer its present perfection.

The great textile inventions outside of England were the cotton gin of Eli Whitney in the United States, to be fully described in a later chapter, and the mechanical pattern loom of Jean Marie Jacquard of France, of more importance to silk than to cotton. As remarkable as was this later machine, it cannot be classed as a truly original creation, since it is very close to a Chinese loom of the Twelfth Century, pictures of which were no doubt known in France, because of the broad interest in l'art Chinois of that time. The invention of roller printing from engraved copper rollers by Thomas Bell has already been mentioned.

The next century in England and America was indeed a period of hectic invention, but it was directed towards the perfection of detail, the enlargement of the earlier types, the speeding up of machinery and the coordination of all the mechanical devices, together with an almost complete elimination of hand processes.

The first machines were largely concerned with the final processes and had to be supported by a great amount of hand work. It was necessary to radically change the human element in relationship to production as well as to modify the machines. The craftsmen of that time fully realized what the machines meant to them in an economic way, and bitterly and not always unwisely opposed them. The tendency in invention was to break down the last resistance of old crafts and to so arrange production that less and less skill was necessary in labor. Consequently lower and lower wages could be forced through the competition of the unskilled with the skilled and, ultimately, children against adults. This change was accompanied with conditions of human misery, beyond belief, nor was it confined to the textile industries. Production in other fields rapidly followed cotton in the use of machinery and the intense specialization of labor functions. James Watt's steam engine vastly aided this condition, and is indeed a vital part of all modern industry. All that fine sense of interclass responsibilities, so sympathetically described by Froude in his essay *Sixteenth Century Englishmen*, was swept aside and its place taken by confused theories of personal liberty, free competition in commodities and in the lives of men and women.

The distressing human conditions, the unquestioned falling away of standards of merchandise, produced under the first generation of the machine, drew a bitter protest from intellectual England, a revision of the electoral system, a vast migration to the colonies and the modern labor union. This period is in truth the beginning of the modern economic system, and the works of Robert Owen, Thomas Carlyle, William Morris, and John Ruskin may again be read with great profit and perhaps a broader understanding. But in their own time they were prophets, crying in vain in the midst of a wilderness of mechanical greed.

It is not to be expected that the United States escaped some share in these evils. But in this country we were not hampered by a rigid ownership of land; and as one class of labor grew dissatisfied, it was possible to replace it with another, until the present population of our great textile towns in New England is composed of a rather dubious mixture of most of the races of the Near East and Southern Europe.

Putting aside consideration of the human element, this period of development shows the English speaking people at their very highest point of efficiency. Endless patents were taken out, there was improvement in practice and theory, almost from day to day. But during this period only three machines deserve mention alongside of the great inventions of the former century. These are the ring frame, invented in 1832, the Jacquard pattern loom, and the Draper loom.

The mule of Crompton, even in its present form, requires an expert mechanic to operate it; besides the movement is intermittent, as the carriage runs out the yarn is spun, as it recedes the yarn is wound upon the spindle, hence only half of the time is spent in actually spinning. The ring, on the other hand, is a continuous motion. For many years it was thought that the mule spun yarns, especially in the finer numbers, were superior to those made on the ring, nor has this idea been entirely abandoned. Yet there is a growing belief that the ring has great possibilities for quality as well as quantity.

The ring frame was a distinctly American machine invented by Mr. Jenks of Pawtucket, Rhode Island, and for many years the British manufacturers scorned it. It was popular in America, however, because it could be operated successfully up to a certain point by unskilled labor and was particularly suited to the introduction in parts of the country where there were no large bodies of skilled labor to draw from.

The Northrop loom with its automatic shuttle feed was invented by a British mechanic, employed in the Draper Loom Works in Hopedale, Massachusetts, in 1898. It is but fair to say that this invention owes much to the encouragement given by the Draper organization to the inventor and the constant improvements made in the details of construction in types of shuttles, etc., since it was first brought out. In the type of loom which preceded the Draper type there was only a single bobbin of weft. The weaver had to stop the loom and replace this with a new one, each time it ran out. Since the loom runs at the rate of one hundred to two hundred picks of weft per minute, it will be seen that in a mill of thousands of looms, there was a considerable loss of valuable time. The Northrop Loom has a battery of weft bobbins and as the bobbin in the shuttle runs out, an automatic device inserts and threads a new one, without stopping the movement of the loom. The principal economy of this loom lies in the fact that with good yarn, a weaver can take care of a greater number of looms than formerly. It is unquestionably one of the most nearly perfect of all automatic machines.

Eighteenth Century: Age of the Machine

As in the case of the ring, this invention was at first disliked in England, but today it has won its place in the British textile mills making plain fabrics; and a factory has been built in the midlands to make the English type of Northrop loom.

As I have suggested, the machine age was not wholly responsible for the destruction of the individualized craft expression of the guild ages. The guild ages were destroyed, once the principle of the division of labor and specialization in phases of production was established. The machine was rather the result of this process than its cause. The fine traditions of the Middle Ages (themselves built perhaps upon the eastern craft arts, brought back by the Crusaders), were destroyed or rather buried by the deluge of merchandise uncontrolled by these traditions, following the discovery of the all water route to the Orient. The merchant rather than the manufacturer was responsible.

All of our artistic losses may, however, easily be reclaimed. Beauty of pattern, richness of color, true æsthetic values are always within the reach of an appreciative audience. Each age and people always get the art they deserve, I had almost written desire.

But the influence of our methods of production, the personal detachment from creation, the loss of individuality, apparently inseparable from this phase of industry, is another matter far more difficult to estimate, control or modify. We may rely on the advancement of taste to restore us some measure of skilled crafts, based upon the fine traditions of ornament, and even a vast improvement in machine-made fabrics. But the problem of educating our factory population to an understanding of their relationship to production and to life in general, to restore to them the fine joy of creative effort, is a problem that we must take up with patient seriousness, before we breed conditions of discontent, impossible of solution through mere economic formulas.

CHAPTER X

COTTON IN THE COLONIES

WITHIN the present boundaries of our Atlantic Seaboard cotton states and in the valley of the Mississippi, cotton was unknown until introduced by the European colonists. There is an account of cotton blankets being objects of trade between the powerful tribes at the mouth of the Mississippi and the Pueblo tribes of New Mexico, but the plant was a stranger to this region, now famous as the world's greatest cotton plantation.

Its first introduction is said to have been due to Spanish agency. In about 1536 Spanish colonists in Florida are reputed to have planted cotton seeds. In this there is nothing improbable, since the Spanish governors had long since made use of native labor, and may have hoped to find the fierce warriors of the mainland as tractable as the gentle inhabitants of the islands. Cotton lint was already an article of trade between the Spanish and Portuguese colonies and Europe. As early as 1570, the cottons of Brazil were regularly sold in the market of Ulm in southern Germany.

At the time of the first settlement of Virginia in 1606, the British East India Company had been formed six years and cotton was well known in England.

Consequently the colonists brought with them cotton seeds, since they had been informed it would grow in their new homes as well as in Italy. It was hoped that supplies of raw materials might be raised for the home market or at least enough for the domestic needs of the colonists.

In the light of the later importance of cotton both in England and America, these earliest beginnings have been magnified in most works on the subject. As a matter of fact there were no prophets at that time who foresaw either England's great industrial need for cotton two centuries later or America's preeminence in raising the fiber. Quite the reverse was the case. Tobacco, Indian corn, lumber and even silk were of greater importance. It is true that taxes in cotton lint were accepted in the Carolinas, but a special law was passed in Virginia, levying a fine of £10 on any colonist owning ten acres of land who did not plant ten mulberry trees. Special bounties were offered in tobacco to encourage silk raising in 1657. A small reeling plant was erected in Savannah in 1732, and in 1759 ten thousand pounds of cocoons were brought there to be thrown in silk thread. On the other hand no cotton was shipped from our colonies to England until late in the Eighteenth Century.

Clothing was one of the chief problems of our early settlers and few indeed could afford the luxury of imported garments or cloth. Encouragement, therefore, was given by Colonial legislatures and governors to the raising of sheep, flax, and to a lesser degree of cotton for home uses. Almost every farm, even as far north as New Jersey and Pennsylvania raised small patches of the fiber. That this supply in time was not adequate is proven by the fact that, even in

COTTON IN THE COLONIES

1—Embroidered bed spreads made in New England by Mary Breed in 1770, showing influence of Calico patterns. *(Page 127)*
Metropolitan Museum of Art.

2—Double cloth blanket woven in geometric pattern by weavers of Dutch descent in the Hudson Valley. *(Page 127)*

3—Spinning wheels used in the Colonies before the introduction of machinery. *(Page 127)*
U. S. National Museum.

4—Detail of embroidered coverlet. *(Page 127)*

5—Blue and white double cloth blanket woven by weaver of Dutch descent in Closter, N. J., 1833. This is a very early example of American Jacquard weaving. *(Page 127)*

PLATE 17

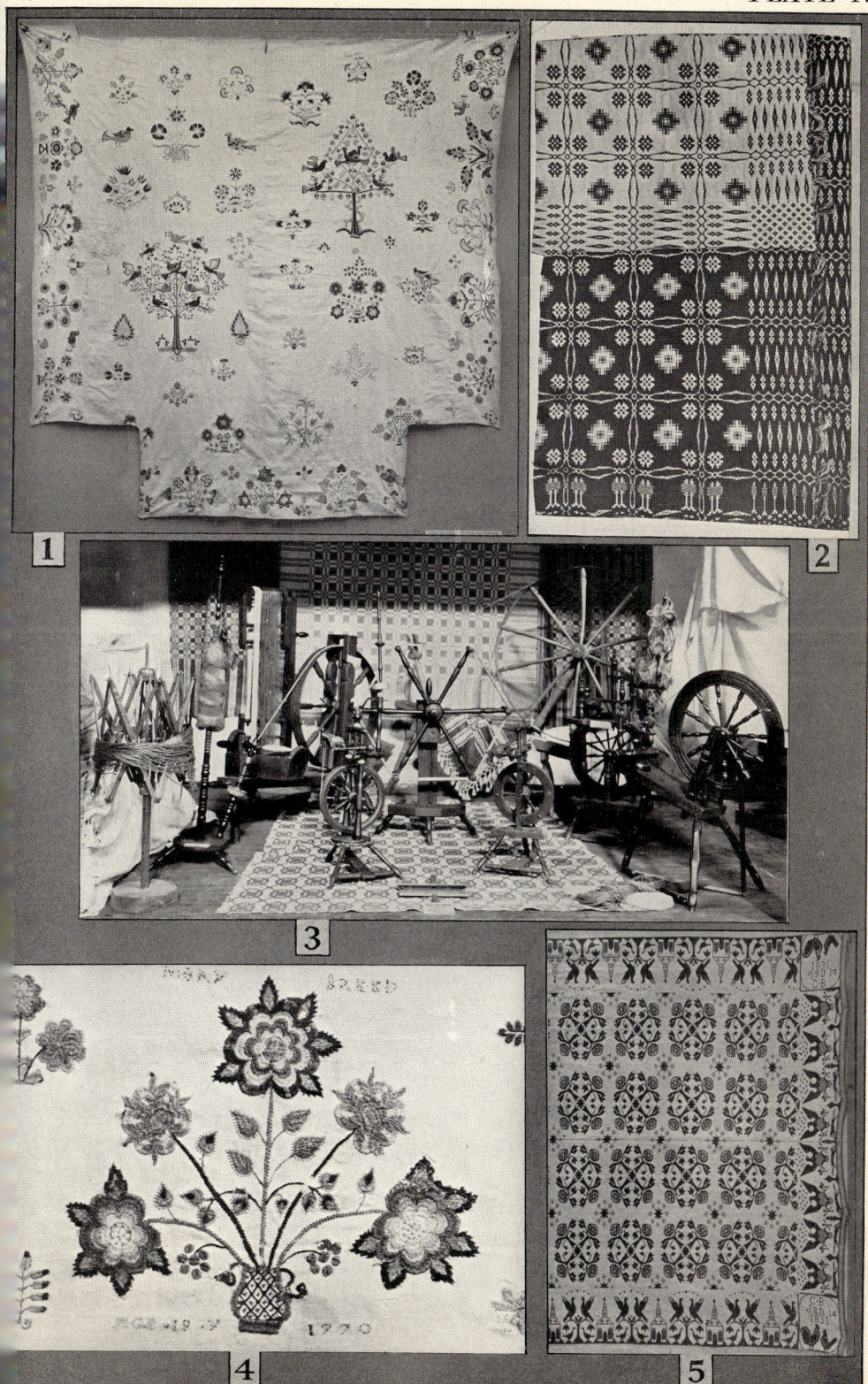

early Colonial times, a considerable supply of raw cotton was imported from the British West Indies and perhaps even from forbidden Spanish ports.

Our ancestors in the main came from the artisan classes of Europe, and among them was naturally a sprinkling of skilled weavers and a certain knowledge of dyeing. Of course, every woman of that day knew how to spin and the old Saxony wheel was a familiar object in every home.

Even today in the older parts of the United States there are evidences of Colonial craftsmanship. In the retarded sections of the Kentucky and Carolina hills, a vestige of this loom-art remains, containing patterns which first originated perhaps in the Far East. It is an art well worthy of encouragement. In New England there was a strong beginning in the early Eighteenth Century of the splendid Elizabethan embroideries. The Dutch of the Hudson Valley produced coverlets of blue and white flax and wool, and may even have used cotton. These coverlets are, in point of texture and design, to be considered as serious expressions of textile art. The Pennsylvania Germans were not only skillful weavers, but introduced block printing in blue and white patterns, a familiar craft in their European homes since the early Fourteenth Century. The prevalence of this art in Pennsylvania may have been one of the reasons which induced Benjamin Franklin, in 1753, to establish a British printer of calico in Philadelphia. There are more or less veracious accounts of attempts of the British soldiers during the Revolution to capture this man, so prejudicial to British trade in this country.

The problem of the Colonial textile workers in color was, of course, very difficult. But these early craftsmen were not lacking in an investigative spirit and turned to the natural products of the soil for their dyes, aided maybe by the example of the Indians. Golden rod, sassafras, gaul berry, horse laurel, iris root, hickory, red oak, and walnut barks are but a few of the substances they used to obtain color. At a very early date, indigo was imported and later raised in this country. Quaint and interesting as these colors and patterns would appear to our eyes today, they did not bear comparison with either the prints of Europe or the beautiful calicoes of India. And we very early begin to hear of commerce in these luxuries, especially when the New England mariners began to make voyages to the East, or to ports where these forbidden commodities were obtainable. An advertisement of Benjamin Franklin, offering a reward for the return of a calico dress stolen from his wife, proves both the existence of this commerce and indicates that fine cotton dresses in rich colors were both highly prized and very rare. The old dock in Salem, Massachusetts, where the tall ships once moored from their voyages to the seven seas, is still, for all its departed grandeur and lost romance, known as India Wharf, as a reminiscence of the time when cargoes of calico were received and exchanged for cargoes of the salt cod so famous in this section of the world.

It is, of course, easy to exaggerate the importance of the Colonial arts, and there is a natural confusion with the products of later times and even with the materials of purely European origin. Infant colonies, struggling for life in a hostile wilderness, are scarcely the proper setting for the gentler arts. Still we must remember that for almost two centuries our growth was stationary, and in our Colonial life, therefore, there was a period of security, stability and leisure, in which the modest crafts might well ripen and develop. With the dawning of the age of machinery and the sudden awakening of that restlessness of spirit which swept our boundaries between two mighty oceans, all of these kindlier, homelier matters were brushed aside, awaiting the return of peaceful days.

Even these modest beginnings of domestic crafts met with small favor in England. Colonies, in the plan of things, were outland centers of population, created to supply the home market with raw material and to be in turn a market for the finished products of the Mother Country. All kinds of manufacturing were, therefore, looked upon with disfavor and amiable Colonial governors were instructed to frown severely on the beginnings of our infant industries. That this policy, however, had little effect on our ancestors is amply attested by the steady increase of the cottage arts and the small shipments of raw cotton to Great Britain.

There is no record of cotton shipped to England from the present limits of the United States until 1764, when eight bags of perhaps one hundred weight each are mentioned. In 1770, three bales of two hundred pounds each were shipped from New York, ten from Charleston, and three barrels of lint seed from South Carolina. When it is recalled that this was well within the age of invention and that the British consumption at the time was over four million pounds annually, it will be seen that our cotton was of little importance to the hungry spindles of Lancashire. At this time seventy per cent. of their wants were supplied from the British West Indies and perhaps from the Spanish colony of Louisiana.

Just after the Revolution, a British frigate captured an American vessel and confiscated ten bales of cotton of two hundred pounds each, on the charge that no such vast amount of lint could have come from this region and must, therefore, represent contraband trade from the West Indies or the Spanish ports of the Caribbean Sea.

Another reason why cottons played so small a part in our early agriculture and commerce lies in the new character of the plant developed under cultivation above the frost area. The first cotton seeds were brought from India, the Levant and the West Indies, even from far off Siam. In these regions the plant is naturally of the perennial variety, growing more like a tree than a shrub. It was quickly discovered that the least touch of frost destroyed the plant and made it necessary to replant each year. This led to confusing mixture of types, and since cotton is notably sensitive to cross fertilization, many curious botanical changes took place. The upland types of cotton of our Atlantic Seaboard States, which became in time the dominant type all through our cotton area, differs from the Indian types and those which prevail in Brazil and the islands of the Spanish Main. The American cotton lint adheres so tenaciously to the seed, that it could not be ginned by the gentler method of the roller gin or churga of India. Consequently all our earlier cottons had to be pulled from the seed by hand, and since labor even of slaves in the New World was scarce and high, this excluded our cottons from world markets and reserved it solely for domestic purposes.

The invention of the cotton gin by Eli Whitney was a characteristic Yankee answer to this enigma. It marks rather the beginning of the modern history of cotton in the United States than a later phase of the first beginnings. This invention did not occur until the first decade of our national existence, and many matters of interest preceded it in the growth of our cotton industry.

CHAPTER XI

THE MACHINE AGE IN THE UNITED STATES AND THE GROWTH OF THE COTTON PLANTATION

THERE is a very natural partisan feeling in most of the accounts of the first beginnings of mechanical production in this country. As a matter of fact the case might be made out as strongly for the South as for the East. Both sections after the Revolution were about equal in wealth, with the South if anything in the lead. In point of energy and knowledge of the outside world, our Revolutionary records indicate that the inhabitants of our southern colonies, particularly Virginia and the Carolinas, were at least equal to Massachusetts. It is probable, as well, that there was a closer intimacy between individuals in our southern colonies and well informed persons in England than in any other section with the possible exception of New York.

It is but reasonable to assume that the South heard the news of the great inventions of the cotton machinery of Lancashire at least as early as New England, and no doubt these inventions were the subject of speculation and perhaps even experiment at a very early period. But it is idle to assume that the true nature of these machines could have been understood by mere description in either the North or the South.

Skillful as the colonists may have been in certain household arts, there is no question that in the British midlands mechanical knowledge and skill were of a more professional character and much wider spread. The machines may have been known of in the South as quickly as in the North, but in neither location was it possible to successfully build or operate them without skilled mechanics from England. There is no doubt, however, that more or less successful attempts were made to secure working models and mechanical drawings in both sections of the United States at about the same time.

In most of the histories of the beginnings of our textile industries many compliments are paid the energy and sagacity of our textile founders, nor are these compliments lacking in point or in truth. Energy and sagacity are indeed the proper words to describe their mental attributes. There seems, however, to have been little if any particular honesty in these early attempts. To be sure the rights of inventors were hardly as yet recognized anywhere, since invention, as we understand the term, was largely a new social and industrial concept. Property in mere mechanical ideas was a little beyond the comprehension of the world at this time. Nor must we forget how alluring were the dreams of sudden wealth that surrounded the modest beginnings of this industrial age. Some parallel perhaps may be seen in the treatment accorded to the inventor Seldon in our own times, by the manufacturers of motors, in their appropriation of his idea of the combustion engine; in the obscurity regarding the early patents on the telephone and the incandescent lamp or even in the dubious treatment of the carefully built up German dye formulas, during and following the Great European War.

There is just enough hypocrisy in each age to deplore the tragic fate of great geniuses of other times, whose patent rights have expired. There is quite enough selfishness, at any time, to take advantage of the carelessness or idealism of inventors to make us just a little careful in passing judgment on the morality of the manufacturers in past ages who appropriated, without proper recognition, the creative ideas of their times.

The first concrete attempt at the international appropriation of machine ideas occurred in Philadelphia in 1786, when a group of local capitalists advertised in England for mechanics willing to break the English laws and escape to America with models or drafts of mechanical spinning devices. A small plant was organized to make cotton yarns, based on information apparently so obtained. From the meagre description of this machine, it appears to have been the James Hargreaves spinning jenny, generally in use in England by 1770.

In the diary of Washington of 1789, that practical Virginian mentions visiting in Beverly, Massachusetts, a small mill, where several cotton threads were spun at a single operation. This enterprise was under the control of a Mr. Cabot, a name not unfamiliar in later cotton history, and the machine was still the Hargreaves spinning jenny. This venture, like the one in Philadelphia, did not prosper. More than merely machine designs and drawings were apparently necessary for success. What was required was a skilled superintendent, familiar with the little expedients and devices of mechanical production, willing to leave England and come to this country with such knowledge as he possessed, and if possible with the contraband drawings and models.

Machine Age in the United States 135

During this period the South was far from an indifferent spectator of events. Hope of encouraging some English mechanic to come to the South was expressed in the resolution of the Safety Committee of Chowan County, North Carolina, on March 4th, 1775, in the following resolution:

"The Committee met at the house of Captain James Sumner and the gentlemen appointed at a former meeting of directors to promote subscriptions for the encouragement of manufactures, informed the committee that the sum of eighty pounds sterling was subscribed by the inhabitants of this country for that laudable purpose."

There is a very interesting note from Stateburg, North Carolina, in 1790, which gives a clear idea of a plant in those days or at leastproves that some one in this part of the world understood the theory of cotton manufacture as it was practised in England:

"A gentleman of great mechanical knowledge and instructed in most of the branches of cotton manufactures in Europe, has already fixed, completed and now at work on the high hills of the Santee, near Stateburg, and which go by water, ginning, carding and slubbing machines, with eighty-four spindles each, and several other useful implements for manufacturing every necessary article in cotton."

In 1790, the first well trained English mechanic, the first man actually informed in the best practices of cotton manufacture, was finally induced by what he had heard of opportunities in America, to seek his fortune in this country.

Samuel Slater first landed in New York and sought to interest local capital in the machines he had carried away in his mind. But there was little interest in New

York at that day in cotton mills. They knew perhaps of the unfortunate experiments in Philadelphia and in Massachusetts, and sought safer ventures for investment. Slater had nothing with him to prove his ability, except the fact that he had been an assistant in Arkwright's mill in Cromford and had worked under Strutt, Arkwright's competent partner. This indifference led New York to miss the doubtful advantage of harboring the early cotton industry.

Slater found in Silas Brown, a well to do Quaker merchant of Providence, R. I., a more sympathetic auditor. With the assistance of this shrewd business man, the young British mechanic was able to build from his memory, some approximation of the machines on which he had been trained in England. Those machines were neither as accurate or as large as the originals nor was his organizing ability perhaps as great as that of his master. Still his cards and spinning frames, preserved in the Smithsonian Museum at Washington, are a great tribute alike to his mechanical ingenuity and the power of his recollection. The concern he founded, but a few years after the Revolution, is still in existence and was within this generation, well known for certain cheap grades of cotton merchandise. The original mill has been preserved in Pawtucket as a memorial museum.

In the first Slater Mill only spinning was performed. The yarns were sold over the eastern states to cottage weavers. Within a few years Slater's example had been followed, and many spinning mills organized in Rhode Island, in much the same manner that they grew up in England following Arkwright's successful experiment.

There is no doubt that these industries were well known to the planters of South Carolina and that many

MACHINE AGE IN THE UNITED STATES

1—The Boston Manufacturing Company, Waltham, Mass., the first mill in the United States to use the power loom. (*Page 146*)
From an old photograph.

2—Slater Mill in Pawtucket, R. I., where yarns were first spun by machinery in 1793. Now a museum. (*Pages 136, 144*)

3—The old Whaler, Charles W. Morgan, in New Bedford Harbor with background of cotton warehouses. (*Page 151*)

4—Spinning machine, patented by Peter Paddleford in 1816. (*Page 122*)
United States National Museum.

5—Mechanical carder built by Samuel Slater for mill in Pawtucket, R. I.
United States National Museum. (*Page 136*)

6—Jenks' ring frame for spinning cotton. (*Page 122*)

7—Spinning frame built by Samuel Slater. (*Page 136*)
United States National Museum.

PLATE 18

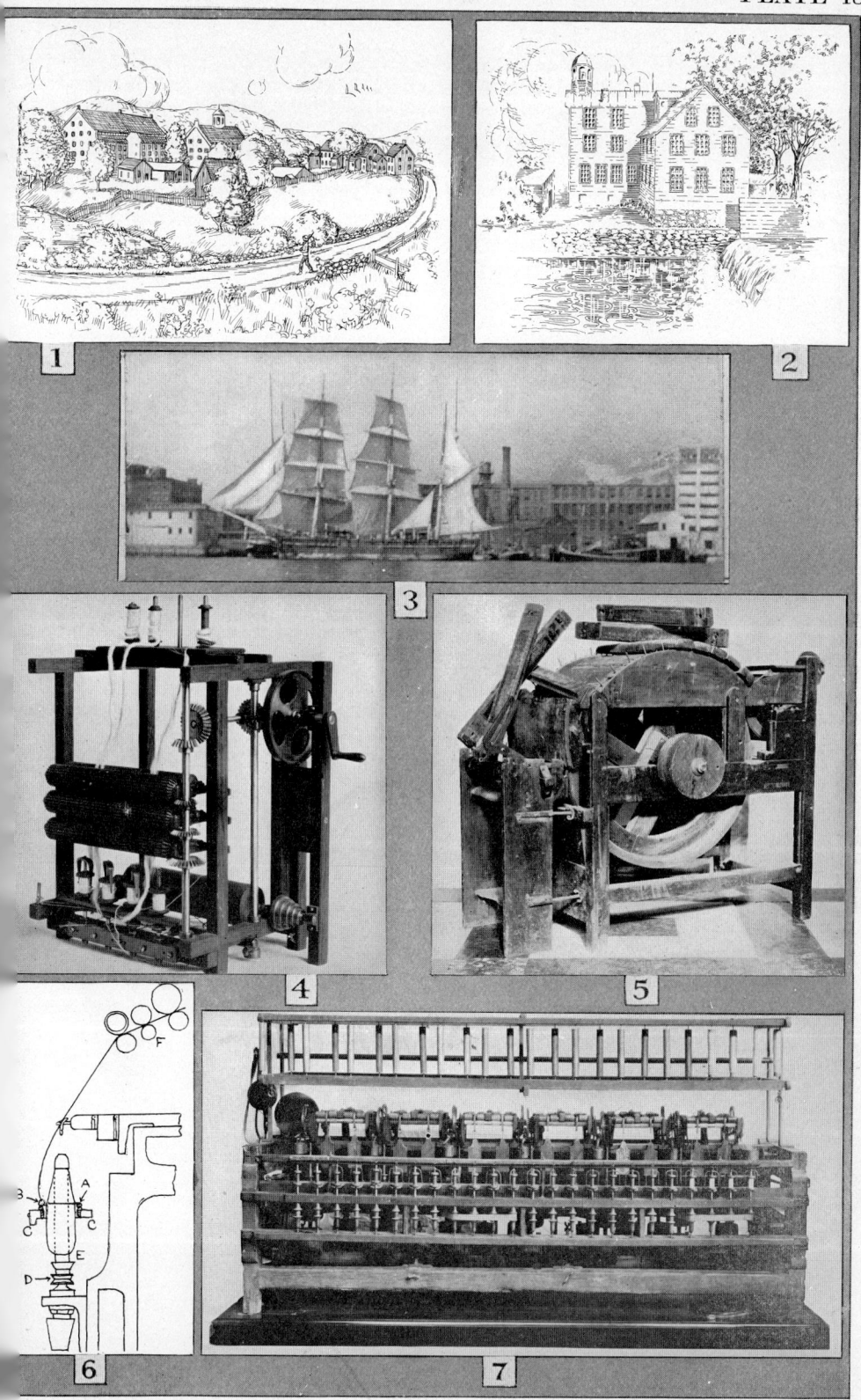

attempts were made, some of them partially successful to emulate this example. Many small mills were started in the South at this period, but the product was largely absorbed on the plantation or in the community in which they were organized.

The South was divided in its counsels as to the desirability of manufacturing as other opportunities for development presented themselves in too alluring a form. The sudden growth of the cotton industry in Great Britain, in the latter part of the Eighteenth Century, had created an immense market for raw cotton. The planters in South Carolina were keenly alive to the possibilities of supplying this market. Here was land that could grow cotton and also a more than rudimentary knowledge of agricultural practice. Here was slave labor to cultivate and pick the crop. England's cotton requirements had risen since Arkwright's and Crompton's inventions and the perfection of roller printing by leaps and bounds. In 1771, England had imported about four million five hundred thousand pounds of cotton lint, seventy per cent. of which came from the British West Indies, the rest from India, the Levant and Brazil. In 1791, imports had risen to over twenty-three million pounds and this limit was not set by England's wants but by the available world supply of cotton. Therefore, between the great and immediate prosperity of the cotton states of the United States lay only the mechanical problem of removing the seeds.

It is small wonder that the great land holders in South Carolina were more deeply interested in this subject than in the production of cotton fabrics by mechanical methods. They were well equipped to supply cotton fiber to English mills, but in no sense equipped to market the finished fabrics even if they could make them. It chanced at this time that young Eli Whitney, graduate of Yale and a citizen of the state of Connecticut, had just accepted a position as manager of the large estates of Mrs. Green, widow of the distinguished Revolutionary general. Whitney had had as good an engineering education as the times afforded and practical experience in mechanics in a factory manufacturing fire arms. He was informed, by the planters of this region, of the difficulty in separating the upland-cotton from its seeds and was easily persuaded to begin a series of experiments to develop a machine to suit the botanical character of the domestic plant. In this he had the full support of the local planters and in 1793, he perfected the original Whitney saw tooth gin, which still gins over ninety per cent. of the American cotton crop.

Briefly, this invention consisted of a box with a flooring of iron grids. Through slits too narrow to permit the seeds to pass, circular saws revolved. Cotton in the seed was placed on the platform and the revolving saw blades tore the lint from the seeds and revolving brushes removed the lint from the teeth. Simple as was this device, it had an immense and immediate influence on the economic life of the South.

Almost at once all idea of cotton manufacturing was abandoned for the sure and immense rewards of cotton planting. This sudden change in the scope of agricultural opportunity immediately altered the attitude towards chattel slavery, since it was soon found that the negroes' immunity to high temperatures made him ideal as a cotton cultivator. Even before this time, the landless white man, of the mechanic and independent labor class, had been at a great disadvantage through competition with slave labor and through social disabilities incident upon his non-ownership of slaves. The cotton bonanza sealed his doom and rapidly forced him to seek the meager security of the mountain country, where cotton could not be cultivated. He will appear later in the history of cotton and take a vital part in its latest phase, but from here on he is swept aside, beyond the currents of events for almost a century of neglect and poverty.

To the slave holding land owners in the cotton belt Whitney's invention meant sudden and vast wealth. We can trace this quite accurately by a brief study of the export and import chart of raw cotton to England during the next generation.

It is estimated that in 1791, there was raised in all the United States only two million pounds of lint. Most of this was absorbed by local needs, only 189,500 pounds being shipped to cotton hungry England. In 1791 this amount had fallen to 138,325 pounds. In the year 1793, the first year of the cotton gin, 487,000 pounds were sent to England, a large proportion of it from South Carolina. In the next year 1794, this amount had more than tripled to 1,601,700 pounds. In 1795, it had again quadrupled to 6,276,300 pounds. For some reason difficult to understand, but perhaps due to the rising demands of the Pawtucket spinners, in 1796, only 3,788,429 pounds went abroad, but in 1798 it rose to 9,360,005 pounds and in 1800 to the amazing total of 17,789,800 pounds. A decade later in 1811, on the eve of war with England, 62,186,081 pounds were exported. In other words, in less than a generation the exports of cotton had increased about fifty fold from the United States alone.

The war with England temporarily reduced shipments almost two-thirds. With the close of hostilities in 1815, the demands of England were so vast that 82,998,747 pounds were exported. It will be recalled that the gallant if unnecessary battle of New Orleans, was won by the dashing Andrew Jackson from behind a rampart of cotton bales. By 1820, the total had risen to 127,860,152 pounds. Nor must it be forgotten in estimating the wealth that poured into the South, that after the founding of the Slater Mills in 1793, New England became of growing importance as a market for raw cotton.

By 1840, there were 2,285,337 cotton power spindles in the United States, 1,599,698 of which were in New England and the rest scattered through the Middle Atlantic and Southern States. In 1860, there were 5,235,727 spindles and in 1870, in spite of the vast disturbances of the Civil War and the first decade of reconstruction, there were 7,132,415 spindles. Estimated in terms of cotton bales of 500 pounds each for domestic consumption, this is as follows:

In 1850 240,000
In 1860 918,926
In 1870 905,243

The fall in these last figures, is accounted for of course by the economic conditions during the early phases of reconstruction.

The South, immediately before the Civil War, was of vital importance to the British cotton industry. England's cotton imports in the year 1859–1860 indicate this in no uncertain terms.

Shipments of cotton into the British Isles in units of bales of 500 pounds each, 1859–60:

Machine Age in the United States 141

United States	2,522,000
Brazil	103,000
West Indies	10,000
East Indies	563,000
Egypt	110,000

Is it any wonder in the light of these amazing figures, that Senator Hammond of South Carolina should have declared in the Senate Chamber at Washington in 1858:
" . . . Would any sane nation make war on cotton? Without firing a gun, without drawing a sword, should they make war on us, we could bring the whole world to our feet. . . . What would happen if no cotton were furnished for three years?"

These words were tragically prophetic, for between 1861 and 1865, the threat was almost made good and southern cotton ports were closed or partially closed by the vigorous naval policy of the Union. There is no doubt that England's economic life tottered under the strain, nor were their wanting advocates in England who desired war between the British nation and the North in consequence. These advocates were strangely among the most cultivated and intelligent of the English people. We should do honor to the memory of the workers in Lancashire, to the very people who suffered most from the lack of cotton, that they put human liberty above their own wants and prevented a tragedy too great for imagination to picture.

But England did not submit tamely to the situation. This has never been her habit. She rimmed the globe with cotton plantations. Wherever cotton could be grown, there her deep purse and far sighted policy planted it, and many of the plantations thus founded, have remained fruitful to this day.

In the last year of the war, 1864–1865, England's cotton import figures, show how successful she was in this venture:

United States......	198,000.......	2,234,000 −
Brazil.............	212,000.......	109,000+
West Indies........	60,000.......	50,000+
East Indies........	1,798,000.......	1,235,000+
Egypt.............	319,000.......	209,000+

In other words by adding the increase of cotton imports from countries other than the United States, we find that the total imports to England during that year were 1,603,000 bales, giving a net loss of 721,000 bales or a little over one-third of her normal pre-war supply. This was a great commercial achievement, and one in which British merchants and statesmen of those tempestuous times may well take pride.

When we compare, however, the prices between these two periods, it will be seen that England paid heavily to keep her spindles and looms at work.

During the year 1859–1860, the average price of cotton ranged between $10\frac{3}{4}$¢ and $11\frac{3}{8}$¢ a pound. In 1865 it rose to the astounding level of $1.82 a pound and on the rumors of peace fell to 43¢ a pound. It must be remembered, however, that during the latter phase of the war American currency had undergone a certain degree of inflation.

I have given here a condensed synopsis of cotton figures since the invention of the Eli Whitney cotton gin, to prove how little real incentive the South had to manufacture cotton goods and every inducement to devote her energy to raising cotton. At the same time, even at her most prosperous times of cotton farming, the germ of manufacturing was kept alive by clear visioned economists, and there were not wanting a few courageous men who endeavored to put theory into practice. I will return to the consideration of mill development in the South, after I have outlined its rise in the East.

CHAPTER XII

MILL BUILDING IN NEW ENGLAND

THE Slater Spinning Mill, 1790, was followed in Rhode Island by others, until a considerable center of mechanical spinning had been built up around Providence and Pawtucket.

New England at this time was a very important maritime center. The great forests gave her an abundance of timber for ships and the fisheries of the Newfoundland banks had built up a hardy class of sailors, never surpassed perhaps in the history of the world. Her ships were in every port of the world, famous alike for their superior sailing qualities and the sagacity of their navigators. New England's daring youth saw the world and, seeing it, learned of its ways and opportunities. Commerce from Ulysses' days to the Yankee clipper ships has been the mother of culture, as of wealth. Had New England adhered to the narrowness of her self-styled Puritans, had she limited her vision to the rim of her granite hills, her history would have been but the drab record of a mediocre, agricultural community. But the blue of the seas was in the clear eyes of her youth, and the rosy dawn in their souls! Hardship and emergency they knew and in self reliance met secure the turn of events.

Mill Building in New England

To this rapidly spreading commerce, the Chinese embargo of Jefferson was a shrewd blow, to be rapidly followed by the War of 1812, which was the reaction from the envy of British shippers. After a gallant resistance, our little navy was either totally destroyed or mewed up in ports, and our merchant ships captured or rotting at the idle wharfs. The blow that might have crushed a lesser people, simply aroused the energy of New England, and turned her to manufacturing. And now the great idea of cotton production began to be taken up in serious interest. Whatever is to be said of later developments in this first home of cotton manufacture, its first growth is a matter of deep pride to all America.

In a former chapter, the invention of the power loom by Cartwright in 1785 has been mentioned. Scottish manufacturers seemed to have been a little quicker to adopt it than the English. They were first to apply power to Crompton's mule and the printing rollers of Thomas Bell, and naturally the loom followed. In an excellent pamphlet by Nathan Appleton in 1858 is an account of the first introduction of power loom weaving in the United States.

The author and Francis C. Lowell met in Edinburgh in 1811 and both became keenly interested in the development of power weaving. Lowell continued his investigation both in Scotland and in Manchester until 1813. It is doubtful if in this undertaking, worthy as it was, he was received by the British mill owners in any too cordial a spirit. Mills in those days were not particularly accessible to the affable stranger. Still Lowell was a young man of fortune and energy with no slight tincture of that sagacity, traditional with the earlier inhabitants of his native commonwealth. He evidently learned enough to cause him to organize in 1812, in Boston, a company with four hundred thousand dollars capital, one hundred thousand dollars of which was fully paid in. The corporation included Appleton, Lowell and Patrick T. Jackson. Their first move was to secure the services of Paul Moody of Amesbury, a skilled mechanic. Moody and Lowell then proceeded to experiment with the power loom in a loft on Broad Street, Boston, Massachusetts.

In the meantime the mill was built at Waltham on the Charles River and cards and spinning machinery fitted up in running order. The first power loom was successfully run in the autumn of 1814, and the venture was at once a success and additional two hundred thousand dollars of capital was added.

It can hardly be said that even in England as yet had the power loom won complete recognition. And it must be admitted that at this stage it was only suited for the weaving of the coarsest fabrics. Its effect, however, upon the cotton industry in New England, may well be imagined; nor is there any reason to question that much of this credit belongs to Lowell. His improvements were highly original and very practical, and he seems to have possessed at once a broad vision and a sure practical sense of values, together with a keen judgment of men. His death, at the early age of forty-two, was a great loss to the budding industry. The organization he founded is still in operation under the name of the Boston Manufacturing Company and the old mill still stands, although an upper story has been added to it. This mill was the first in America in which every process from raw cotton to a finished cloth was carried on.

Nathan Appleton, the author of this valuable

Mill Building in New England

pamphlet, was a practical man of affairs, and confined himself to an accurate description of the types of cloth first made in Waltham. For this I am deeply grateful. It was a heavy sheeting of number fourteen yarn, thirty-seven inches in width, forty-four picks to the inch, and I assume about forty or forty-two in the warp and weighing about three yards to the pound.

During the War of 1812, the absence of all foreign goods from our markets made cotton manufacturers very prosperous indeed. As the manufacturers of more recent times, they attributed this benefit from an ill wind to their own superior sagacity rather than its principal cause. The peace of 1815 swiftly disillusioned them and brought ruin as rapidly as hostilities had brought a vicarious prosperity.

Here we come upon the first instance in a long series, where manufacturers looked to legislation for aid, rather than to their own ingenuity and vigor. Pawtucket was a desert of forced idleness. Only a few spindles were running in the Slater Mills to make yarn for the local home weavers. The manufacturers braved the dangers of stagecoaches and appeared before Congress, petitioning for a tariff high enough to force back the miraculously vanished profits.

Lowell, confident because his own mill could run in competition with those of England and perhaps not particularly sensitive to the misfortunes of his neighbors who did not have power looms, counseled moderation. He persuaded Lowndes and Calhoun to support a minimum tax of $6\frac{1}{4}$¢ the square yard. This undoubtedly gave him adequate protection and also left him free from any immediate danger of domestic competition. Could we ever again come happily on so simple a tariff as suggested by Lowell, it would save us from many a perilous calculation. He advised the manufacturers of Rhode Island to seek in power looms a cure for the ills of trade. Whether or not Lowell supplied the want he himself had created I do not know. It may be that he did sell some of his rivals power looms or technical skill to produce them. There is no doubt that his Job-like comfort bore fruit, as is indicated by the rapid shrinkage in the price of cloth in the following table. These prices refer to the type of sheeting above referred to as made in the Boston Manufacturing Company.

1816	30¢
1819	21¢
1826	13¢
1829	8½¢
1843	6½¢

I am further indebted to Mr. Appleton for an account of the founding of the great modern textile city of Lowell, Mass.

A group, of which Appleton and the invaluable Paul Moody were members, organized a company to take over the water power site of the Merrimack River for the Merrimack Manufacturing Company. This company was incorporated February 5, 1822. In August, seventy-five thousand dollars was paid to the old Boston Manufacturing Company for its machinery patterns and patent rights, and also to release the essential Mr. Moody for his new duties. In September, 1823, the first wheel turned, and in 1825, the first dividend of one hundred dollars a share was paid. Up to 1855, in spite of occasional hard times, an average dividend of 12½% was paid. Nor should it be forgotten that stock watering was by no means an un-

Mill Building in New England

known art at this time. In addition to this, the rapidly growing value of real estate and water power were not overlooked. Manufacturing was then a most profitable venture in all its aspects.

The Merrimack Company, if not the first to actually print calicoes with rollers, was the first to have organized a company in America for this purpose. There were mysterious and fruitful visits to England and attempts to discover the secrets of Sir Robert Peel's great success in the business. English engravers of copper cylinders were induced to settle in America, as well as chemists and dyers. The mill founders of those days were men of vision and energy, able and willing to press their adventures to a successful issue.

From here on the story of Lowell is the story not so much of cotton as of water power. The Hamilton Company was founded in 1825 with six hundred thousand dollars capital, later increased to a million two hundred thousand. In 1828 the Appleton and the Lowell Companies were organized. In the depression of 1829, Amos and Abbott Lawrence were induced to take a large share in the water power holding company, and established the Suffolk, Tremont and Lawrence Companies in 1830. The Boott followed in 1835, and the Massachusetts in 1839. These companies involved capital of twelve million dollars.

"In November 1824, it was voted to petition the Legislature to set off a part of Chelmsford as a separate township. The town of Lowell was incorporated in 1826. It was a matter of some difficulty to fix upon a name for it. I met Mr. Boott one day, when he said to me that the committee of the Legislature were ready to report the bill. It only remained to fill the blank with the name. He said he considered the question narrowed down to two, Lowell or Derby. I said to him, 'then Lowell by all means,' and Lowell it was."

This is, of course, merely the outline history of one cotton town, but one that is still of great importance to the industry and typical of the staple branch of manufacturing in New England.

There are many brief records of mills unsuccessfully started in many other sections, or of small unimportant successful ventures. In 1789 sail cloth mills were established in Haverhill, Salem, Nantucket and Exeter. In 1790 the first checks and ginghams were woven on hand looms and in Rhode Island by 1810 there were a dozen or more spinning plants, inspired by the success of Slater's venture. It is estimated that in 1810 there were sixty-two plants in the United States, operating 32,000 power spindles.

It was not to be expected in New England that the great success of the Pawtucket, Waltham and Lowell manufacturers would pass unnoticed. New England was then at the high tide of her energy, before the South and the Middle West began to attract her youth into other fields of endeavor. Wherever there was adequate water power, there came shrewd men, intent on founding cotton mills.

The Great Falls Manufacturing Company was organized in 1823, Amoskeage in 1831, Laconia Mills in Biddeford, Maine, in 1845, Pepperill, in 1850, and Pacific in Lawrence, Mass., in 1854.

Salem, the ancient seaport town, famous for her witches and her clipper ships, when she saw her ocean trade diminishing in favor of Boston and New York, organized the Naumkeag Steam Cotton Mills in 1839. It did not get in operation, however, until 1847. This is the first instance of the use of other power than that furnished by the turbulent rivers of New England to run looms and spindles, and marks the dawn of a new era in industry.

All of the concerns I have mentioned are still in existence, have grown immensely during the last few years and have in the main been prosperous. They are typical examples of the type of cotton mill on which a large measure of the early prosperity of New England depended. I am well aware that many other mills might be profitably included in this list. For many years, in fact up to modern times, building in New England has been a constant experience, but this is not a chronicle of organization, but the history of a fiber, and I can only touch the distinct phases and not completely exhaust each detail.

The beginning of the fine yarn industry was in the little city of New Bedford. No city in all the world during the first half of the Nineteenth Century was so famous for the whaling industry as this little town on Buzzards Bay. Her sturdy vessels were known in the far ports of all the strange oceans. In China and the South Sea Islands the United States was regarded as a suburb of New Bedford, since across the stern of all vessels flying the Stars and Stripes appeared the name of New Bedford. In this city today, there is a most charming little museum, containing a half sized, fully equipped model of the square rigged whaling bark, the *Lagoda*, and a complete assortment of all the implements used in this romantic combination of big game hunting and commerce, together with the quaint and beautiful objects brought back by the hardy mariners from the many strange ports at which they touched. Along the steep and narrow streets near the harbor, there are ship chandlers and sail lofts, and the other accompaniments of seafaring towns; nor has the ancient city entirely lost its flavor of adventure, for still come to dock a few battered schooners and old square riggers, sad ghosts of a vigorous yesterday.

In 1847 while the whaling industry was at its zenith, a few capitalists in this wealthy little seaport town were induced to organize the Wamsutta Mill to make fine shirtings. At this time the staple business in cotton fabrics was pretty well controlled by the great corporation mills in other sections of New England. To compete with these thoroughly established institutions was wisely regarded as a hazardous investment, and it was obvious to these men that the rapidly increasing wealth of America had created a market for better qualities of cotton goods.

It is rather curious in the light of recent events, that the man who first conceived the idea of a cotton mill in New Bedford should have been an employee in a small cotton mill in the South, owned by Dwight Perry, a resident of Fairhaven, just across the bay. Thomas Bennett, Jr., was, however, a former resident of New Bedford and was ambitious to run a mill of his own based upon his experience in a southern mill. He appealed to Joseph Grinnell, a member of Congress and a well-to-do citizen of New Bedford. Mr. Grinnell was at first strongly inclined to locate the mill in the South, but was finally persuaded, rather against his will, to make the venture in his native city. With the aid of David Whitman, a mill engineer of Rhode Island, the mill was finally put in operation in 1849 with fifteen thousand spindles and two hundred looms. The capital investment was about $160,000.

This venture was a success from the start. There

were at this time about ten thousand men engaged in the whaling industry, and this meant a large number of women available for mill labor. In 1854 two additional mills were erected, bringing the total spindlage up to 45,000 with an adequate complement of looms. In 1907 this corporation was capitalized at over three million dollars, had 228,000 spindles and 4,300 looms and employed 2,100 operators.

Few people outside of the original stockholders apparently cared to invest in cotton mills in this city for almost a generation. Whaling was still too profitable a venture, and most of the local capitalists were concerned in financing these journeys and disposing of the oil, bone, and ambergrease. The losses of the Civil War due to the Confederate privateers, together with a series of tragic disasters in the Arctic Ocean and the dawn of competition with mineral oil, changed the local sentiment, as a list of mill building during the next twenty-five years indicates. At first, however, the progress was slow, gradually gathering momentum, until at the end, almost each year at least one and sometimes two new mills were added.

Wamsutta Mills	cloth	1847
Potomska Mills	cloth	1871
Acushnet Mills	cloth	1881
Grinnell Mfg. Co	cloth	1882
New Bedford Mfg. Co	yarn	1882
City Mfg. Co	yarn	1888
Bennett Mfg. Co	yarn	1889
Howland Mfg. Co	yarn	1889
Hathaway Mfg. Co	cloth	1889
Pierce Mfg. Co	cloth	1892
Columbia Spinning Co	yarn	1892
Bristol Mfg. Co	cloth	1892

Rotch Spinning Co	yarn	1892
Dartmouth Mfg. Co	cloth	1895
Whitman Mills	cloth	1895
Soule Mills	cloth	1901
Butler Mills	cloth	1902
Gosnold Mills Co	cloth	1902
Manomet Mills	yarn	1903
Kilburn Mills	yarn	1904
Page Mfg. Co	cloth	1906
Nonquit Spinning Co	yarn	1906
Taber Mill	yarn	1906
Quansett Spinning Co	yarn	1906

Many other mills have been added since this list. The Sharp Spinning Mill of 200,000 yarn spindles, which has within the year added a thousand looms, the Booth Mfg. Co., and the Nashawena Manufacturing Company are typical of this later period.

Today in this city there are in the neighborhood of 3,500,000 fine yarn spindles and 50,000 looms. Nothing but fine cotton goods are manufactured here and usually they are light in character and always of combed yarns. There are no better equipped mills in the world than the mills of New Bedford.

Curiously enough, this largest group of quality cotton mills are generally unknown to the public with the single exception of the oldest mill, the Wamsutta. They do not finish, dye, bleach or print their own product, but sell it in an unfinished state to the merchant converters of New York and other large cities. Fabrics are sold in the "gray," shipped to different finishing plants and converted into merchandise, according to the inclination of the merchant converters. These are sold in turn to the department stores, the jobbers and the costume manufacturers. In other

Mill Building in New England

words, the final determination of the character of the product, its style, its quality, etc., is made by factors outside rather than inside the mill.

The labor population of this little city is of the most interesting character. Successively Yankee, British, French Canadian, and Portuguese have predominated. Certain of the Slavic and Mediterranean peoples of other New England towns are not strongly represented here. New Bedford requires skilled labor, and to train skilled workers even in our modern industries is a slow process. Except in managerial and in technical positions, the American and British labor element have moved on to more fruitful fields of opportunity, although a sprinkling of both still remain. French Canadians are present in fair numbers, but the Portuguese predominate.

There are two kinds of Portuguese, the white and the dusky natives from the little Island of Brava in the Cape Verdes. In rich, dark, chocolate tones, these can not be distinguished, except by the expert, from our own negroes. There is, however, a fine difference in temperament and it is wiser to keep the distinction well in mind. They are industrious, thrifty and under reasonable management docile, but they distinctly do not regard themselves as in any sense politically or socially inferior. As islanders, they are naturally excellent mariners, belonging, however, rather to the heroic age of the sail than to steam. The few whalers that still use this little port, have Brava crews and in many cases Brava or Portuguese navigators. Anyone familiar with the type of vessel, its equipment and general seaworthiness, still employed in the whaling trade, will be quick to realize that any body of men, willing to act as sailors in these vessels, must have a deep confidence in their own ability to contend, almost unaided, with the forces of nature.

Whaling has a rather interesting modern relationship to the problems of cotton manufacturing. It is one of the sources of foreign labor supply, that seems to be exempt from the inquisitiveness of all government interference. Whenever a whaler finds itself in the neighborhood of the Cape Verde Islands and has had little success with the wary denizens of the deep, it becomes an impromptu passenger vessel. There are always a few adventurous and covetous souls in Brava, who sigh for the El Dorado of New Bedford. Such of these as possess either credit with the skipper or the necessary funds, board the tiny, sea-worn craft and come to America with rather less comfort and often greater danger than that famous mariner, who in 1492 touched at the Canary Islands in order to mend a broken rudder.

In the days of the infamous slave trade, these islands were used by the slavers and the escaped Africans added in time a deep colura madura tone to the complexion of the natives; but they are aggressively Portuguese, none the less, and usually of good intelligence and more often literate than their white compatriots. Besides the occasional trips of the belated whalers, they have their own distinct fleet of vessels, plying between Buzzards Bay and the Islands. Condemned Gloucester fishing schooners, square rigged barks, that have fallen in the toils of the excise officers and been auctioned off, in fact any sailing craft, passed the fastidious requirements of our own seamen, are purchased by these daring mariners and, after a few rudimentary alterations, become what are known in the local parlance as "Brava Packets." There is a firm belief in New

Mill Building in New England

Bedford, shared by all of true citizenship, that a real Brava would think nothing of going home in a Cape Cod cat boat. When times are dull in New Bedford, these boats trade among the islands and when word comes of steady work and good pay, they set their threadbare, dingy wings to the boisterous air of the Atlantic and change the capstan bar and the halyard for the spinning throstle and the loom. Since all passengers and sailors alike come as crew, such as choose may desert when in port. The immigration officials pay very little attention to the matter.

On a recent visit to New Bedford, a dilapidated Portuguese gun boat, about the size of a lighthouse tender, visited Fall River and New Bedford, and in both towns the mills shut down for the simple reason that nobody appeared to work, since practically the entire population gave themselves over to celebrating the great event.

I have chosen two typical New England towns, famous for cotton production in early days, and briefly, perhaps too briefly, sketched their growth. In a general way their history is duplicated in other cotton centers, although no actual parallel on so large a scale exists for New Bedford.

I have omitted intentionally any general discussion of labor conditions, reserving this for a future consideration.

For the moment I will review briefly the beginning of the merchandise systems still in operation.

When the Boston Manufacturing Company, in 1815, first produced power woven cotton goods, the problem of selling became at once acute. Attempts to dispose of this merchandise through an importing house were not successful. At this time there was only one place in Boston where domestic goods were sold. This was a shop in Cornhill kept by Mrs. Isaac Bowers. At this time only a single loom was running in Waltham, yet even this modest product Mrs. Bowers found it impossible to sell. Everybody praised the goods but nobody bought them. The stigma of "domestic" came into existence on cotton fabrics at the very birth of the power industry. The next shipment of goods was consigned to B. C. Ward & Company at Mr. Appleton's suggestion and was eventually disposed of through an auctioneer.

"That it was so well suited to the public demand, was a matter of accident," continues Mr. Appleton. "At that time it was supposed no quantity of cottons could be sold without being bleached; and the idea was to imitate the yard-wide goods of India, with which the country was then largely supplied. Mr. Lowell informed me that he would be satisfied with twenty-five cents the yard for the goods, although the nominal price was higher. I soon found a purchaser in Mr. Forsaith, an auctioneer, who sold them at auction at once at something over thirty cents. We continued to sell them at auction with little variation in the price. This circumstance led to B. C. Ward & Co. becoming permanently the selling agents. In the first instance I found an interesting and agreeable occupation in paying attention to the sales, and made up the first account with a charge of one per cent. commission, not as an adequate mercantile commission, but satisfactory under the circumstances. This rate of commission was continued, and finally became the established rate under the great increase of manufacture. Thus, what was at the commencement rather unreasonably low, became, when the amount of annual sale concentrated in single houses amounted to millions of dollars, a desirable and profitable business."

Since this firm, or at least Mr. Appleton personally, was a stockholder in the mill in Waltham, the smallness of the commission can be easily explained. So begins the practise of commission houses handling the product of the corporation mills. As more and more mills were built and enlarged, as the trade grew and grew, this small percentage changed into vast sums and commission houses became very rich indeed. Not only do they own stock in mills, but the mills own stock in commission houses, and there is a complicated, intricate, interlocking ownership between rival commission houses machinery manufacturers, printing plants, bleacheries and textile banks, that would take a Philadelphia lawyer to even partially untangle.

The goods usually handled through commission houses today are the so-called staples,—muslins, ginghams, denims, heavy shirtings, sheets and pillow cases, bed tickings, etc. These are sold largely to the jobbers in bulk lots under more or less well known brands, and by these re-sold to the thousands of retail stores all over America. Within recent years there has been a tendency to sell directly to certain of the large retailers and in some instances to the garment makers. The ultra conservativeness of the jobber has led to this change in policy, since it is a deterrent to any progressive policy on the part of the mill.

The manufacturing of ready-to-wear garments, compared to the cotton industry, is a very recent economic development. Twenty years ago, this remarkable industry was in its first crude phases of growth and was little regarded by the long established and wealthy cotton mills. This system of selling has largely isolated the commission houses from that close observation of the development in fashions and styles, that is the basis of successful commerce in fabrics in America to-day. The garment industry very swiftly outgrew the mills' idea of designs and quality, and the relationship between these industries is rather slight. The mill executives determined the types of fabrics largely by their mechanical convenience. It was simpler to manufacture a few supposedly well established and desirable types of goods, than to make a highly varied line. So the mill put designing on a purely mechanical basis, and devoted their entire attention to the steady run of their looms and spindles, with little regard to the shifting demands of the public.

For a long time, there were great areas in this country where conditions of life were sufficiently archaic to create a desire for the fabrics that had been popular in our grandmothers' times. There was also a great increase in population, which offered a certain vicarious market to almost any kind of a product, so long as it was cheap enough. If periods of depression came, so did periods of inflation, and there was always the alurement of the cotton exchanges, and the discreet issues of stock in good years to make the absence of profits through poor styling pass unnoticed.

The rapid growth of our silk industry, forced to work in close accord with our wholesale dressmakers, was attributed by the sagacious cotton mill owners to the superior, natural merits of the silk fiber. They looked upon the silk industry with considerable tolerance, since it offered them a cheap and convenient method of getting designs for their prints; and also gave an added value to the mercerizing process, which made an imitation of silk that fooled nobody except

Mill Building in New England

the cotton man. (I am well aware that good mercerizing increases the strength of cotton yarns when properly and thoroughly done, and when the type of cotton used in the yarn and the twist of the yarn is correct.)

The somewhat contradictory fact that fine and expensive cotton goods, well constructed and in excellent patterns, were yearly growing in our import figures, was dismissed by the cotton mill men and the jobbers on the ground of the natural perversity of an unappreciative public. The remedy lay in praying to a usually sympathetic Congress for still higher tariff. The whims of the non-technical public have never been permitted to disturb the serious counsels of this long established industry.

The attitude of the English manufacturer of cotton goods differed little if any from that of his American cousins. A greater skill in labor communities, a more diversified market for products tended to make certain types of English goods a little finer in technical construction than the American, but neither England nor America have recognized the absolute commercial value of good style, which is but another name for good taste. This was and still is in a large measure regarded as a matter beneath the august consideration of so serious a body as the board of directors of a cotton mill. So the craft organizations and designers of France year by year seek out and earn our most profitable cotton trade. Today French cottons, often hand woven, dyed and printed, or made in little mechanical plants closely approximating hand craft, sell freely for several times the price of domestic goods; and the curious condition exists of a very dull market for staple goods and a splendid sale for fabrics of charm in pattern and genius in construction.

The parallel between the preference of the public as expressed for the beautiful merchandise of France, as against the excellent but uninteresting textures of American and English origin, and the early history of the calico trade with India, is most significant. Nor are there wanting excellent business men in America and England to whom the prohibitory statute of 1700 would be regarded as an economic benefit. Few of these economists have any other method of defense indeed, in spite of the fact that during the terrible years of the War, the entire producing energies of the French people left their market open all over the world, to be captured by ingenuity and energy. It will be remembered perhaps by some future historian, writing in happier days of our industry, that the fine yarn mills of New England, at this particular time, devoted their energies to the chimerical effort to secure the purse of Fortunatus, through letting their looms stay idle and spinning yarn for automobile tire fabrics of a quality which has since been determined to be far in excess of the requirements.

CHAPTER XIII

THE SOUTH

AS I have pointed out, the early development of mechanical production of cotton goods in the South was interrupted by the immense advance in cotton cultivation, due to the invention of the Whitney Gin. This arrestment in industry did not come suddenly, however, but only when the world market for raw cotton was assured beyond question. As late as 1810, the manufactured products of Carolina and Virginia exceeded in value those of all New England. Some few mills in the South, however, continued to produce yarns and fabrics, many of them in locations since occupied by the great mills of the present day. But they were local affairs, in many instances merely plantation mills and in some few cases even run by slave labor. In *The Rise of Cotton Mills in the South*, by Broadus Mitchell, the distinction between the economic development from 1810 on in the East and the South is very clearly and ably drawn. The statement of a southern banker, quoted by this author, clearly indicates that the distinction once established existed down the beginning of the last generation.

"The mills built after the War (Civil War) were not the result of pre-bellum mills. This is trying to ascribe one cause for a condition which probably had many causes. The industrial awakening in the South was a natural reaction from the War and Reconstruction. Before the war there was first the domestic industry proper. Then came such small mills about Winston-Salem as Cedar Falls and Franklinsville. These little mills were themselves, however, hardly more than domestic manufactures. When, after the War, competition came from the North, and from the larger southern mills, the little mills which had operated before and had survived the War lost their advantage, which consisted in their possession of the local field. . . . The ante-bellum domestic-factory system did not produce the post-bellum mills."

The following statistics taken from the Census Reports are illuminative:

	Census	Plants	Capital	Operators	Spindles	Bales Consumption
Southern States	1840	248	4,331,078	6,642	180,927	
	1850	166	7,256,056	10,043		78,140
New England	1840	674	34,931,399	46,834	1,497,394	
	1850	564	53,832,430	61,893		430,603

These figures, incomplete as they unfortunately are, prove beyond question that the East far out-distanced the South in manufacturing a decade before the Civil War. Yet there were not wanting prophets of an industrial tomorrow for the South even in the earliest times. Among these William Gregg stands out prominently, alike for his enthusiasm as for his clarity of vision. In 1845, he said:

"Since the discovery that cotton would mature in South Carolina she has reaped a golden harvest; but it is feared it has proven a curse rather than a blessing. Let us begin at once, before it is too late, to bring about a change in our industrial pursuits . . . let croakers against enterprise be silenced. . . . Even Mr. Calhoun, our great oracle . . . is against us in this matter; he will tell you that no mechanical enterprise can succeed in South Carolina . . . that to thrive in cotton spinning one should go to Rhode Island. . . ."

The South had not only an economic but a social disinclination towards manufacturing, and steam power was actually forbidden in Charleston, S. C., and skilled white labor from the North was not encouraged to settle in the South because it was felt that both of these factors might be out of sympathy with the institution of slavery.

There were approximately three distinct periods of cotton mills in the South, between 1790 and 1833. I have mentioned the interest in British machinery in Colonial times and the experiments made, together with the establishment of a number of small plantation mills. After 1812 a few northern mill men came into the South and established small plants, because it was impossible then to ship yarn to the southern markets because of the English blockades. Between 1820 and 1833 the tariff differences with the North encouraged a little mill building in an attempt to meet the argument of a protective tariff by actual physical competition from the South. But the first growth of the South as a manufacturing center begins in 1840. Only after this date do the mills begin to lose their first domestic character and assume the organization of factories comparable in many ways to those of the North and East. This growth was again retarded by the political activities of the cotton cultivators directed towards defending the institution of slavery against the political attacks of the North.

After the War of the Rebellion followed that dreary decade of reconstruction, when the South bowed under the crushing weight of political servitude and all her energy was devoted to the task of merely living until the evil days should pass.

So for the great upbuilding of the modern industrial South, a date earlier than 1870 or 1880 is impossible to set.

In Charlotte, N. C., in Greenville, S. C., in Spartanburg, N. C., and La Grange, Ga., have grown up great cotton manufacturing towns, rivaling those of New England in size and economic importance. Today nearly half the spindles and looms of America are south of the Mason-Dixon Line and the tide is just at its full flood. In actual weight and yardage of yarn and fabric the balance already lies in favor of the South. This is because coarser types that run more quickly still predominate in the South, and because lack of labor restrictions permits each operator to run more looms and spindles.

A decade from now may easily witness an even greater dominance of the South in the manufacture of staple cotton goods, including even the finer grades.

A decade ago "southern goods" was a term of reproach. This has passed away and many high grade fine yarn mills are located in the South and Greenville, S. C., has become the serious rival of New Bedford, Mass. Bleacheries, dye houses, printing and finishing plants are slowly following, and even some of the manufacturers of textile machinery are building plants in these districts.

In the manufacture of machinery, the South is,

THE SOUTH

1—Model of Eli Whitney cotton gin, invented in 1793. *(Pages 119, 138)*
United States National Museum.

2—Primitive method of spinning and weaving, surviving in parts of the south. *(Pages 127, 165)*
United States National Museum.

3—An almost perfect cotton field with the skyline of Dallas, Texas, in the background. *(Page 220)*
Courtesy of the Dallas Chamber of Commerce.

4—Mechanical cotton picker experimenting before delegates at the First International Cotton Conference in 1919. *(Pages 216, 224)*

5—Bales of cotton being moved along the tracks in Memphis, Tenn., fine cotton warehouse. *(Page 220)*

6—Idealized cotton fiber showing tapering from butt to tip and spinning convolutions. *(Page 214)*
Drawn under the direction of James McDowell.

7—Sea Island cotton under high magnification. *(Page 214)*
James McDowell.

8—Cotton boat "Whisper" on Wolf River at Memphis Levee. *(Page 220)*

9—Photograph illustrating damage done by boll weevil. The central boll was unaffected by the pest. *(Page 223)*
James McDowell.

10—Egyptian cotton under magnification. *(Page 214)*
James McDowell.

11—Arizona pima cotton grown in Salt River Valley, under magnification. *(Page 214)*
James McDowell.

12—Mississippi Valley cotton under magnification. *(Page 214)*
James McDowell.

13—"The America," the largest Mississippi River steamboat carrying cotton, at the New Orleans wharf. *(Page 220)*

PLATE 19

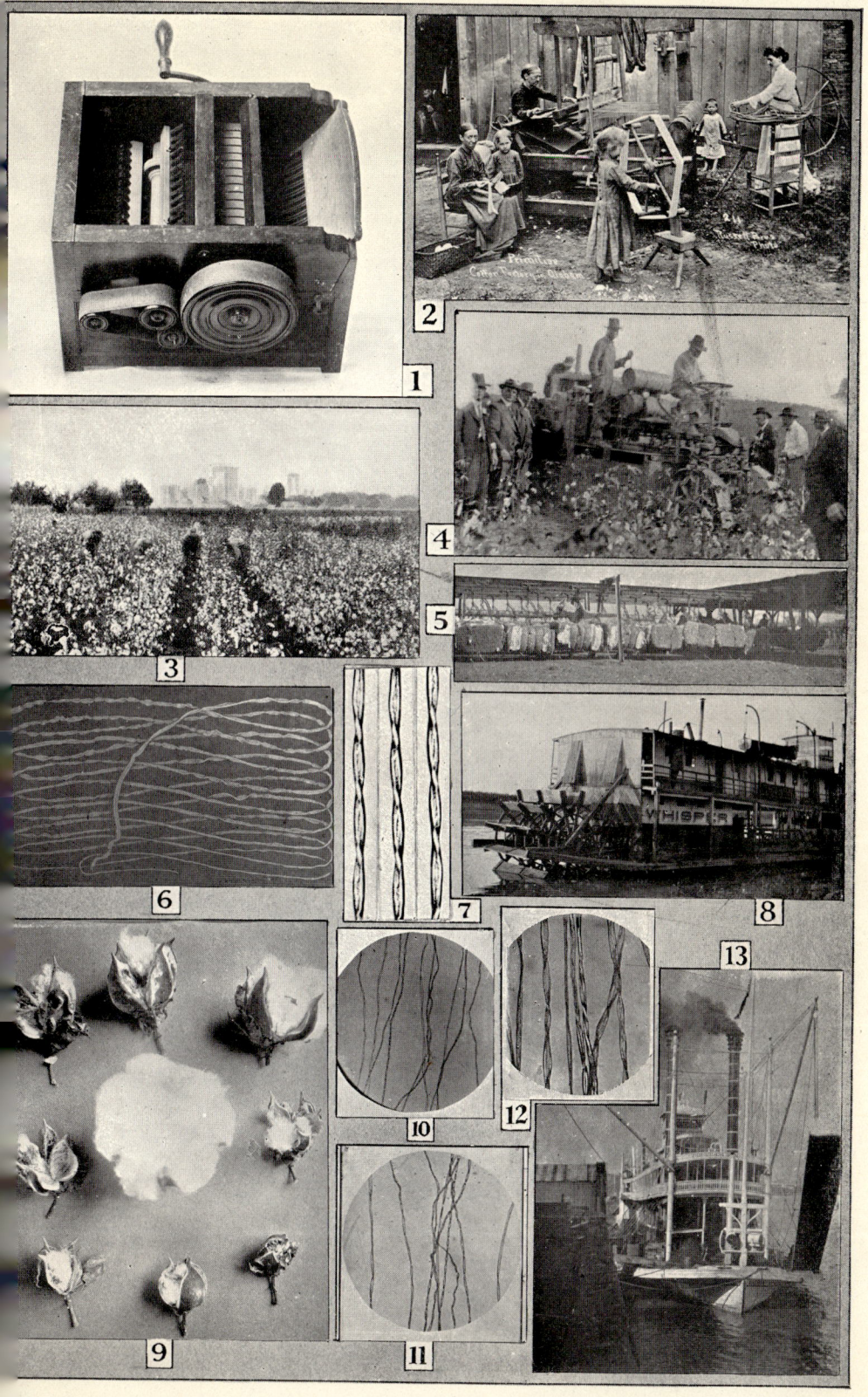

however, still far behind the North and East. In 1919 in Massachusetts seventeen thousand four hundred and thirteen mechanics were employed in the manufacture of machinery, and Massachusetts produced 54.7% of all textile machinery made in America, leading all the country by a wide margin in this phase of the textile industry. North Carolina, the nearest southern State, was ranked ninth and employed three hundred and five mechanics and produced only 1% of the machinery. South Carolina one hundred and thirty-four mechanics and .04% was ranked eleventh. All the New England States outranked all the southern States in the production of textile machinery.

In the production of yarns and fabrics, the story is quite different. In 1919 Massachusetts produced 253,295,403 pounds of cloth and South Carolina 268,270,258 pounds. In yarn the cotton States produced in 1919 57.6% of the total spun in America, and the New England States 37%. Only ten years previous in 1909, the cotton States produced only 23% of the yarn and the New England States 73%. In 1919 New England had 17,542,926 spindles, the cotton States 14,568,272 spindles. In 1904 New England had 13,911,241 spindles, and the cotton States 7,495,905.

These figures are highly significant when compared to the development in machinery and textile finishing plants. The South has increased in all branches of the industry in which relatively unskilled labor could be used. There is no reason to believe she will not follow in other branches of the industry where local conditions are favorable.

The South owes her first great development and present strong position in the textile industries largely to her own energy. In the early phases of her latest development she had little help from Northern or from foreign capital. Her rise begins while still the passions and prejudices of the Civil War were powerful factors, while the evils of reconstruction were everywhere apparent, and while mill building in the North and the East was at its height. At this time southern mill stocks were not looked upon as good investments, and southern fabrics and yarns were regarded as inferior in quality.

If the southern textile industry received scant encouragement from outside, at home it did not lack for staunch supporters. The terrible years of reconstruction had burned the lesson deeply into all minds that independence could best be won through the conversion of its principal raw material into yarns and fabrics. To this doctrine the South has religiously adhered up to the present day.

It must be fairly stated that at first every sound principle of conservative finance was against the plan. The North was in no sense antagonistic, merely tolerantly indifferent. Mills in the South lacking the sage guidance of long experience, often inadequately capitalized, without the support of friendly interlocking directorates, selling houses and far from the sources of trained labor, seemed helpless. On the face of things all this was true enough. The difference lay entirely in the point of view.

In the East cotton mill building had become a matter of business to be judged entirely on the probabilities of immediate and safe profits. In the South it was a matter of local patriotism, a gallant effort to meet a crushing situation and turn apparent disaster into prosperity.

The Atlanta exposition in 1861 marks one of the crucial turning points in the new industrial South.

"To exhibit to the southern people and to visitors from America and Europe the different processes in the manufacture of cotton from the boll to the complete fabric, and by the friction of competition ascertain the best methods and find the best machinery. We, the people of the South, should embrace every opportunity which will bring us intelligent and interested observers of our industrial condition, resources and aptitudes. We have in the midst of us the raw material of a magnificent prosperity. We lack knowledge, population and capital. These may be slowly accumulated in the course of years, or they may be rapidly, by well directed efforts to obtain them from beyond our own borders. We advocate the latter plan."

It is judged that over two million dollars worth of machinery was sold at this exposition and it was a great inspiration for many local ventures.

In every town through the South, there were large groups of idle men and women with no opportunity for employment. If the South were to live, to rise above her present level, indeed if she were not to sink to still lower depths, work must be found for these idle hands. The leaders of the South did not wait for any saving miracle, nor delay until capital was plentiful and markets arranged for in advance. She called upon her people, each according to their means and beyond for help. The record of the actual losses sustained in unsuccessful ventures might have appalled any people, must surely have discouraged any people looking only for safe investment. It is not too much to describe the first decade of building as a political venture, almost a crusade.

The words of Hammond, of Piedmont, may sound strange to ears accustomed to more familiar and less eloquent language used in industrial prospectuses. None the less, this was the language of the South of that period and no one may question its sincerity.

"Cooperation . . . is the very spirit of democracy—concern for the common good, not only feeling that I am my brother's keeper, but more I am my brother's brother. We have at last awakened to the fact that the whole is greater than the part. Too often heretofore we have thought of a social class, a segment of interests. . . . But a better day is dawning when we are alike embracing in our affections the whole people, the lowly no less than the lofty. . . ."

The Charleston Manufacturing Company printed in its first bid for stock subscriptions the following message:

"The advantages, direct and incidental, accruing to every citizen of Charleston from this industry about to be started in our city, are so manifest that those who have inaugurated the enterprise have every reason to feel confident of a ready response to the call for capital and of abundant success."

In estimating the sincerity of these sentiments, we must bear in mind the destitution all through the South at this period. Four years of bitterly contested warfare, a complete breakdown of commerce, agriculture and industry, the change from slave to free labor, together with the venality and stupidity of the reconstruction period, had placed this section of the country in a position where it could only be saved through the energy and self-devotion of its own people.

The building of the mills in the South had, in the end, two somewhat interested friends in the East and the North. The commission houses were interested in obtaining merchandise for sale, to compete with the goods from the more firmly established and better financed mills of New England. The machinery people were anxious to increase their markets and to keep their output at a steady level. There is no question that the commission houses, in many instances, drove hard bargains with the under-capitalized southern mills. As a general rule, the mills were built under conditions which exhausted all local capital and made them the easy victims of the sudden vicissitudes of the market. There are stories of mills built out of the meager savings of communities caught in the grip of modern Shylocks that are pitifully tragic.

Usually larger commissions were charged for selling such merchandise and southern mills had less control in disposing of their product than eastern mills. Many of the difficulties, however, were the natural results of under-capitalization, lack of understanding of market conditions and distance from the scene of operations. It is but fair to write that there were many commission houses who took an active and personal and thoroughly honorable interest in the mills they represented and who helped in a most constructive manner in their upbuilding.

In the main, the machinery people were entirely friendly. Their anxiety to broaden their market and keep their plants at the full peak of production is natural to understand. In many instances they took stock in part payment, sometimes amounting to 40% and 60% of their indebtedness, and many a mill in the South today owes its start to the humanity and intelligence of our great machine shop executives.

By 1895 the southern mills had safely passed beyond

the first stages of sectional enthusiasm and desperation into a more or less clear, economic position. They were recognized as the producers of cotton staples of a somewhat inferior quality, their goods were handled by commission houses and jobbers and were sold to the garment factories, making the lower types of men's and women's clothes. Capital had begun to flow in large amounts and eastern technical men to seek careers in southern mill towns. The struggle was over.

At about this time the difference in labor in the southern and eastern mills became apparent. In the eastern mills the American mill hand had already become a myth. He had been pushed out, or rather had voluntarily left the mills and been replaced by the cheapest kind of foreign labor. There had been many waves of migration, some of it voluntary and much of it induced by agents provocateurs of the mills. The editors of certain foreign newspapers in New England were often so employed. Eastern mill towns had been turned into the most sordid labor markets the history of modern democracy can show.

The social and economic distinctions between mill operators, executives and owners were complete and rigid. This story reflects little credit on anyone. Before it could be controlled, it had brought New England almost to the verge of disaster.

In the South the mill owner and his workers were from the first neighbors, speaking the same language, believing in the same political and social ideals, and facing a common danger. If wages were low, so were profits. If hours were long, they were equally so for all concerned. If women and children worked alongside of each other, so had they in the scrubby little home farms. Work in the mill meant comfort, comparative comfort, a measure of security and the hope of the future. Outside of the mills lay chaos. At first the idle men and women in the town were employed, but before long the country people and the distant communities in the hills began to come in to work. The mills were the Mecca of all who had only their labor to sell in exchange for a means of livelihood.

I am no apologist for our industrial system, nor do I see in textile or any other type of industry's general attitude towards humanity over-much to praise, and much indeed to condemn.

But anyone who does not believe that the first decade of cotton mills in the South was a direct benefit and aid to the people of the South, must read history with their prejudices, not with their intelligence. A northern newspaper man, just after the Civil War, wrote:

"Whether the North Carolina 'dirt eater,' or the South Carolina 'sand hiller,' or the Georgia 'cracker' is lowest in the scale of human existence, would be difficult to say. The ordinary plantation negro seemed to me, when I first saw him in any numbers, at the very bottom of not only probabilities, but also possibilities, so far as they affect human relations; but these specimens of the white race must have reached a yet lower depth of squalid and beastly wretchedness."

There can be no two honest opinions on the question of benefit conferred by the cotton mills on the landless white in the South. The experiment, however, of plunging an outdoors people into mills, and harnessing them to a sedentary occupation, the sudden contact between farmers with the traditions of the Eighteenth and perhaps Seventeenth Century methods and modern industrialism, has had many undesirable consequences.

When the first enthusiasm wore off and the mills were face to face with economic pressure and opportunity, there was unpleasant exploitation in many sections.

The record of child labor is not particularly good, the record of female labor conditions in wages and hours is not particularly attractive. As a matter of fact, the cotton mill was never, under any circumstances or among any people, a proper place for children. I can not say, however, that during these periods there is much to choose between the South and the North or between America or England. In all of these countries industrial conditions were too new, and it was natural, perhaps, that many grievous errors should occur. In addition to this the South had to meet the competition of the skilled labor of the world and could only do so at first by confining her efforts to fabrics within her technical powers, and by paying lower wages and working longer hours. To this the East made economic answer by colonizing in her mills every nationality where economic and social evils had created reservoirs of unemployed and secondary cheap labor. To bring the inhabitants of villages of southern Russia, Italy, Greece and Asia Minor into the full rigors of a New England winter, unprepared and unknowing, to lure them by specious promises of rewards in terms of their own currency, was a far more grievous crime against society than the South's first attempt to win economic independence by united action.

Shortly before the great War, the last phase of cotton mill building began in the South. Shrewd mill operators in the East had come to the full realization that certain types of fabrics could be more profitably manufactured in the South than in the East and that southern labor, under proper superintendence and with proper equipment, was the equal of any in the world. Land was cheap, relatively cheap, labor was extremely reasonable in its attitude toward capital and capital was plentiful, since in each community was a surplus of wealth, due to other manufacturing interests, commerce and banking, ready to match the investments of the East in well run mills.

Outside the great mill centers where production has concentrated, every little cotton city or town either has its local mill or ardently prays for one. It is impossible to visit one of these little cities without meeting enthusiastic advocates, all anxious to prove the advantages of their own particular location for the establishment of a mill. Even Texas has the beginnings of a textile organization of its own. There are several small mills, making coarse cotton cloths, sheetings, etc., and about one hundred and seventy-five thousand spindles. Last year Texas produced over a quarter of the entire crop of cotton in the United States, and the smallness of her spindlage compared to her agricultural preeminence is regarded by enthusiastic Texans as a reproach.

The arguments most often advanced why mills should locate in the South are freedom from labor unionism, the relatively low cost of living and nearness to raw material. The fact that the South is a natural market for staple cotton, because of its large agricultural population and climate, is advanced as a supplementary argument.

So far as raw material is concerned, this argument scarcely applies, especially after the mills have reached any degree of organization and size. Few large mills can be run wholly by local supplies of cotton. To make cotton yarns economically and properly requires a careful and judicious mixture of types of cotton. The great variations from season to season in cotton farms make it improbable that any one region, accessible to the mills, could constantly supply them with cotton. One season it will be too good, another season too poor. As a matter of fact, today South Carolina is actually importing Egyptian cotton.

Transportation of cotton to mills, that do not have access to all water transportations, is very little less in the South than it would be in the East. As a matter of fact, there is very little difference in cost between southern mill towns situated in the upland country of Georgia and the Carolinas, and the cost of shipment to the mills of Lancashire.

It is true at the present, that the unions have nothing like the hold in the South that they have in the East or in England, but at the same time, the town of Charlotte, N. C., is already thoroughly unionized and the unions have made some little headway in Greenville. The old prejudice against foreigners (that is people who come from outside of any town in the South) makes it impossible for the average type of eastern mill organizer to successfully operate in a southern mill village or city. The people, however, have themselves an almost perfect genius for organization and with concentration of mills in cities, the rapid distinction being made between the operating and owning classes, the economic vicissitudes of over-production and depression, must bring in time a labor situation, just as was created in the East.

There is no question, however, that so far as the production of staple cotton goods is concerned the advantage today is all with the South. The lack of skill in the operators that once existed has almost vanished. The mill superintendents and executives in the South, together with the excellent technical and extension schools, have worked wonders in the mechanical aptitudes of these people. There is the incontestible advantage of training a homogeneous people, speaking the same language and in general subject to the same economic and social ideals, and the end can not fail to be more highly trained mill communities in the South than in the East.

The statistics over a generation, which of course do not define the subtle differences in the character of the machines, still show the rate of industrial equalization between New England and the South very clearly.

1890

	Spindles	Bales of Cotton Consumed
New England	10,934,297	1,502,177
South	1,570,288	538,895

Total Consumption of Bales in United States, 2,518,409.

1922

	Spindles	Bales of Cotton Consumed
New England	17,938,805	1,853,153
South	15,906,165	3,977,847

Total Cotton Crop in the United States (short crop), 8,360,153.

The year 1907 is regarded as the banner year in the production of cotton in the United States. The government receipts give 10,882,385 bales. It was generally believed at that time that this estimate was very conservative.

CHAPTER XIV

RESEARCH

THE need for research in any industry is too obvious at this late date to require any defense. Most of our great modern industrial undertakings make annual provisions for this essential work, and regard their laboratories as the very soul of their enterprise. Funds are also set aside by groups for scientific investigation in our technical universities and a few progressive associations are attacking their general problems through corporate activities.

In the textile industry there are the beginnings of such developments, but nowhere has the work progressed to the serious proportion it has assumed in other industries. Only within the last decade have the trained technicians from our textile schools been recognized in the industry in any marked degree. And even today the scholastic preparation in these schools does not compare to that given to our engineers in other fields.

Our textile schools have, however, done wonders and compare favorably with those in any part of the world, but they have not gone far enough as yet in building up their courses of study and in connecting these with other fields of knowledge to be compared with our great engineering universities. This is a great pity, since few industries have so many intricate problems or conduct operations, including the selection of raw materials, with so little reference to modern scientific methods of procedure. There is still too much trial and error in our methods of practise, and too many vital questions are left unsolved or accepted on hearsay tradition without searching investigation. The entire field is open to new conclusions and methods based upon a careful, constant and thorough investigation of facts.

Cotton is a seed hair intended in nature to assist in the distribution of seeds just like the fairy parachutes of the milk weed. Like the milk weed fiber, it is an almost pure vegetable cellulose, but here the comparison stops. Cotton fiber has a distinct physical quality which makes it possible to spin it into strong and even yarns. From the root of the seed hair to the sealed tip runs a tube filled with oil during the ripening of the fiber. When maturity is reached, this oil retreats into the seed and the tube collapses. This causes the walls to take on curious, half-formed, spiral turns. Under magnification these look like uncompleted springs. In spinning the fibers are laid parallel to each other through the operation of carding into the soft roll or lap, then gradually drawn out and twisted through a series of processes until the final yarn is spun. These spirals caused by the collapsing walls of the cotton fiber are the adhesive element that makes it possible to spin them.

In the final analysis, therefore, it is the number of spiral turns of the fiber to the inch that determines the evenness and strength of the yarn. This is why the longer staple cottons, other things being equal, are always used to spin the finest grades of yarn. There is, however, another consideration often overlooked and one which led to a serious difficulty in the fine cotton area of Arizona established in the last decade around the Roosevelt Dam in the Salt River Valley.

It is obvious that the greater the number of fibers possible to be included in a cross section of yarn, the greater the strength of the yarn will be, because the strength of the yarn is determined by the degree of friction created in the contact between the spiral twists in the fiber. Mere length of fiber, however desirable, when other qualities are equal, is not the sole spinning determinant. It is the relationship between the number of twists, the length of the fiber, and the number of fibers that can be included in the cross section of any given yarn, that are important. The spinning quality expressed in strength and evenness of yarn depends, therefore, on the number of contact points of the spirals in the cotton fibers in the cross section of yarn. Hence, the finer the diameter of the cotton fiber, the more suitable it is for spinning even, strong yarns. The number of fibers in a cross section of yarn can be counted and the sum of their individual strength determined. This has, however, very little to do with the strength of the yarn, because when the yarn breaks, not more than ten per cent. of the fibers are fractured; the rest is the slipping of the fibers past each other. Spinning quality may, therefore, be rated as follows: spirality—length and fineness of diameter of fiber.

The cotton buyer's job is not only to select good, spinnable cotton fiber, but just the combination of types that will yield the desired results with the least waste and at the lowest cost. He must keep in mind as well the types of machinery in his mill and the most practical adjustments in reference to quality and production.

DEVELOPMENT OF COTTON YARN

The process of spinning is a succession of operations; first paralleling the fiber, next forming them into a soft bolus or sliver and gradually drawing them out and twisting them around each other.

1—Card Sliver:—Soft untwisted rope of parallel fibers. Cotton has previously been ginned to remove seed and passed through the opener in the mill to separate the fibers from the bale pressure, and to dust out the coarser foreign matter.

2—Sliver as it appears after passing through the drawing rollers, to even up the inequalities of the card slivers and slightly attenuate the mass.

3—Slubber Roving:—First process in which the draft and twist are combined.

4—Intermediate Roving:—Second process of draft and twist.

5—Jack Roving:—Process in which twist exceeds draft.

6—Finished yarn.

PLATE 20

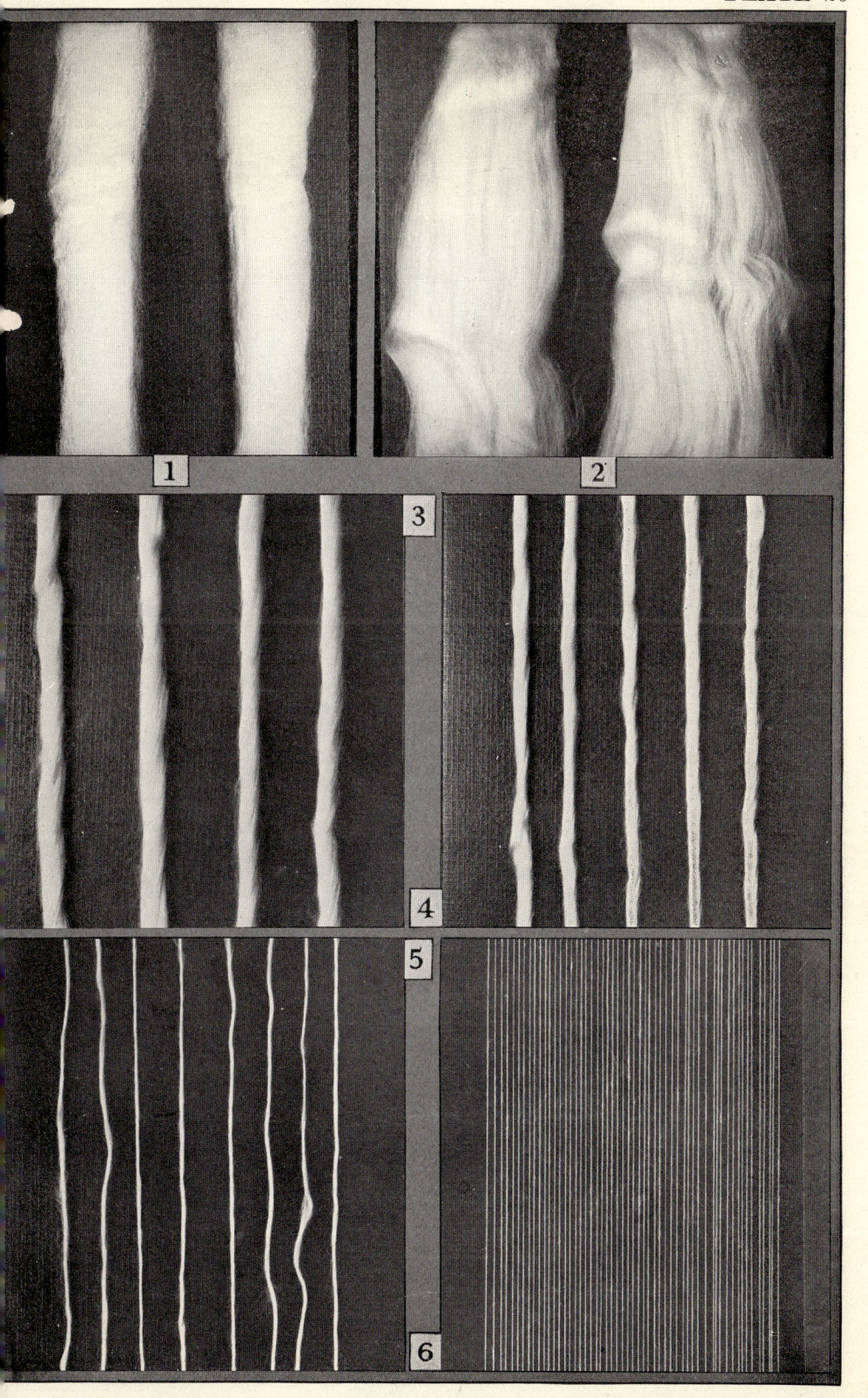

1

2

3

4

5

6

Cotton is a highly variable raw material. Each region has its own types determined by the seeds commonly used, the fertility of the soil, the care in cultivation and general climatic conditions. Cotton is very sensitive to cross fertilization through insects, principally bees, and it is consequently almost useless for a man to attempt to grow finer types of cotton than his neighbors are planting, because through cross fertilization, the types in any given area will be quickly merged.

Over one hundred and fifty grades are recognized by the expert buyer. And here again I must make a very careful distinction between gambling in cotton futures and the outright purchase of fiber to be converted into cloth and yarn. In one case grading is merely an arrangement for convenience in settling gambling debts, and the other is a more or less accurate guide to determine the purchase of raw materials for definite manufacturing purposes, where quality, ease of production and economy are the salient points.

In addition to these already confusing circumstances there is a great variation in fiber quality from season to season on the same farms. This makes the establishment of any empirical standards impossible. In spite of this obvious difficulty, most buying of cotton fiber is done on the personal judgment of experienced men who have become, through years of observation, extremely accurate in their judgments. There is no question that this experience is of the utmost value.

Experienced cotton buyers can take a handful of cotton between their fingers, study it under a good light, and come surprisingly close to a determination of its character and value. First they study it to see how much broken leaf, sticks or other forms of foreign matter the mass contains. Next its degree of whiteness, and finally they pull it apart in little tufts, laying the fibers parallel to determine the length of the fiber and the average evenness of length. In pulling the lint apart, two things are observable. In the first place, they can determine whether the cotton has fully matured, whether it has the proper degree of spirality by listening to the little crisp sound it makes as the twists are pulled by each other. This they call the "sing" of the cotton. Dead or immature cotton does not "sing." They watch carefully to see also that no particles of the fiber fly up in the air during this operation, as this would indicate dead or over-ripe fiber that would be too brittle for spinning.

As I have said, it is amazing how close the conclusions of two independent experts will be on the same lot of cotton, and this is proof of the sensitive skill acquired through long observation and experience.

Obviously, however, such a method is open to serious dangers. Hundreds of millions of dollars are spent each year for raw cotton, and the security of invested capital, considerably over a billion dollars, depends in no small measure on the accurate judgment of the cotton buyers. Some additional precautions are, therefore, desirable before cotton buying can be safely adjudged as a scientific procedure.

All of the physical qualities of cotton fibers are so minute and delicate that they are observable only under a powerful microscope. One of the most thoroughly trained cotton buyers in America has added to a splendid experience a very complete scientific outfit, and his experience over a number of years of cotton buying is proof that his judgment is sound.

James McDowell, cotton buyer and technical

expert for the Sharp and Hamilton Mills in New England, and the Brighton Mill in New Jersey, has come the nearest to working out an exact scientific method of cotton buying. His method of procedure is roughly as follows:

Each season as the first bolls of cotton open on the farms from which he purchased fiber, the previous season, specimens are forwarded to his laboratory. He tests these specimens by the usual methods I have described, and finally examines them under a powerful microscope and photographs them for record. If he detects in the early blooms any serious faults, he can then eliminate some particular farm, or indeed entire area. On the other hand, if he discovers some particular merit in the cottons of some region specially favored during that season by a delicate balance of rain and sunshine, he can direct his buying in accordance with his discoveries.

In addition to his microscope and micro-photographic outfit, he has miniature bleaching and dyeing keirs, delicate strength tests for fiber, yarn and fabric. When he has finally selected the types he wishes to buy in accordance with the orders placed in his mills for yarns or fabrics, he then sends to each mill the proper cotton with instructions as to the mixtures of types to be used in each particular product. Every lot of cotton he receives he tests through the microscope and makes a record of his finding. In the event of any cotton being below the requirements, he eliminates this or averages it up with some stronger type.

In the event of any claims from one of his customers for goods that are not up to standard, he sends to the mill for sample pulls of the bales of cotton used in this lot. He photographs yarn or fabric under dispute and makes careful photographs of the cottons used and compares them with his former record. In this way it is possible for him to determine whether or not his mill is at fault and if so to rectify the mistake.

A very large claim was once made against this mill by a manufacturer of cotton ribbons, who had found that in the light shades little specks appeared in the finished fabric. This manufacturer was buying yarn from three spinners and had divided the claim equally between them as he had divided his orders for yarns. Mr. McDowell insisted that he make a length of ribbon entirely from the Sharp Mill cotton and dye this. He found that the Sharp Mill cotton dyed perfectly. Mr. McDowell's explanation of this fact was that in selecting cottons through the microscope, it was easily possible to detect any undue percentage of unripe fiber which will not dye. The other yarns were made from types of cotton that contained more unripe fiber than the buyers had any idea. It may be a matter of interest to state that the types of cotton used in the Sharp Mill at that particular time cost less than the types used by the other concerns.

In a period of over ten years, during which time Mr. McDowell has purchased over 80,000 bales of cotton a year, he has never had a single bale brought to arbitration, and this should be sufficient proof that his system is sound.

It is evident, therefore, that along this line of research there are great possibilities. The scientific selection of the proper types of cotton for specific purposes is still in its infancy and large economies may easily be effected in mills by the aid of the microscope and the camera, to say nothing of more highly specialized implements that might easily be devised.

If such research were properly coordinated with the Growers Associations, and planters learned the types of fibers best suited to their farms and to the mills, a great deal of waste could be eliminated, costs of yarns and fabrics actually lowered, even if the prices of cotton fiber were higher. It is needless to say that such research can not be conducted either entirely by disassociated scientific men or by the average cotton buyer who depends entirely upon his experience. It is necessary to coordinate all of these efforts and to bring into such a movement the sincere men who are working along agricultural lines with the different farm associations in the cotton area.

Many substances enter into the manufacture of cotton goods besides cotton. The list is practically endless. Moss from Iceland, powerful acids, oils, starches and solvents of different kinds, are all factors. There is besides all these, the great problem of color and the intricate chemistry of dyeing. Here there is the greatest need for cooperative research.

In general the mill superintendents and dyers have only what is called a practical knowledge of chemistry. On the other hand the dye and the chemical industries are founded on accurate trained and constant researches in the higher realms of this intricate science. The communication of ideas, therefore, between these groups is extremely difficult, and there is much unnecessary loss due to lack of understanding and sympathy.

Unless the efforts of the cotton buyer, the spinning and weaving master can be coordinated with the chemist in dyeing, bleaching and general finishing, the best results can never be obtained. There is annually a great wastage in actual material and a loss of quality which might be remedied through intelligent efforts in these fields.

The dye industry in the United States is largely a result of the War. Before the opening of hostilities, America as well as England depended almost entirely on Germany for coloring matter. The entire industry of the synthetic production of color began with the discoveries of Perkins in England late in the last century. It was neglected by the English chemists and taken over and developed through German patience and industry. The result was that within a generation of the discovery of the first mauve shade developed from coal tar, the dye makers of the Rhine Valley held the entire world in their control in the fields of color.

Germany held the monopoly in dyes and through the Cartel system effectively stifled all foreign competition. The production of chemical dyes from coal tar distillation is a very expensive and highly delicate undertaking. There are innumerable by-products, all of which must be developed before dyes can be manufactured with profit. Besides this, the by-products of one process become the raw material or reaction agents for another. Until, therefore, complete units could be organized and methods of self preservation devised, the foreign units could very easily be eliminated by German competition. The Cartel system permitted all German dye producers to combine in foreign trade while competing at home.

If, therefore, a dye plant were started in a foreign country, the particular product it made was reduced below cost, and other products it did not make, including very often its raw material, raised to cover the difference. The opening of a dye plant in any particular country was usually the signal for Germany to dump in that country immense quantities of the chemical made to discourage the local manufacturers. In this way, up to the opening of hostilities, Germany held the textile trade of the entire world completely at her mercy. There is even some evidence on which to base the belief that in certain types of color and fabrics the time was not far distant when Germany might have so regulated prices of dyes that it would have been cheaper to buy the finished product in Germany. At least before the War great quantities of British yarns and fabrics were shipped from England to be dyed and returned to England to sell.

Between the manufacturers of dyes and explosives there is a very close chemical relationship. There are certain stages of the processes when the basic chemicals are identical. In a general way it is almost impossible to be sure of adequate supplies of explosives in the event of war without at least the skeleton of a dye industry in time of peace. The great shells which battered down the gallant defense of Liège and Antwerp rudely awakened the world to this chemical truth. There were terrible days when it seemed as though Germany held the world at her mercy because of her generation of devotion to the sinister science of chemistry.

It will some day be told in detail how English and American chemists met this great issue. Treasures vast and unmeasured were poured into our chemical industries, and men of skill, imagination and no little devotion soberly addressed themselves to these problems. Before the War ended, the gunners of the Allies had just as deadly charges as their enemies, and as a by-product in England and America, dye industries had reached a point where they could in most instances compete with the former monopolist.

There is little doubt that this relationship between dyes and explosives and the confusion between national security and the ordinary uses of commerce were, and still are, made much of by the proponents of British and American dye industries in the securing of highly beneficial tariff legislations after the War. It is a great question where self interest ends and sincere solicitude for the public welfare begins. As a general rule, both in America and England textile interests are in favor of almost free trade in dyes. They regard the chemicals essential to giving color to cloth and yarn as raw material, and this point of view is naturally distinct from that of the producer of the dyes.

The question is not an easy one to settle.

It is difficult to understand why one industry should have anything like special favors accorded them. No man in charity can have two opinions on the question of war of any kind. Few men in a democracy can escape a kind of aversion to specially protected, gigantic industries which from their very nature are almost beyond government or social control. I am sensitive, deeply sensitive, to all of these considerations and freely admit their weight. Yet it is impossible to escape the fact that any nation inadequately equipped in industrial chemical units will undergo immense industrial limitations in peace and be open to disaster in the event of war.

Today America is almost independent in regard to dyes. It is true that certain of the more intricate products used in small quantities we must still import, and it is true that the German chemical interests have attempted to reenter and preempt their old market, and have made rather interesting price concessions to this end.

American dye makers are like our weave masters in that they do not pay much attention to any product that cannot be produced in bulk. In some respects there has been a falling off in research since the War, with the general readjustment that has taken place in other industries. But the truth is that no matter how high a price we may have paid for it we need never again be at the mercy of any nation for either dyes or explosives.

Among the great variety of dyes which were known indeed before the War, but which have been brought into great prominence since, one group is particularly interesting to the cotton manufacturer and to the public. These are the so-called vat dyes, developed in most instances from an anthracene base and called "vat dyes" because the application of color is in vats under immense pressure. These colors on cotton are absolutely fast to sunlight and washing and while in some respects they do not give the rich, beautiful tones of other dye substances, this is not because of any inherent lack of beauty in the chemicals themselves, or in the method through which they have been applied, but because the chemist seldom possesses a nice understanding of color values.

No event, since the invention of Whitney's Cotton Gin, has had greater significance for the textile industries than the developments in the artificial silk industry during the last decade. In a sense, the exorbitant prices of cocoon silk during the War, when this lovely fiber sold for over eighteen dollars a pound, the rapid improvement in knitting machinery, and the high demand for novelty cloths, were responsible for the last great development in this field. But a broader interpretation may be found in the culmination of chemical and physical research in the industry itself.

They had arrived at a place, after patient research and courageous expenditure of capital, that entitled them to the full consideration of all phases of the textile industries.

The history of artificial silk curiously begins at about the same period as that of the machines in England. In 1734 the great French chemist, Rèaumur, made experiments in specially concocted varnishes which were driven through minute holes in sheet iron and precipitated in brittle and unusable threads.

In 1858 Andemars, the Swedish chemist, partially perfected artificial filaments through dissolving the inner bark of the mulberry tree in alcohol and ether.

The great genius of the industry, however, was Count Hillaire de Chardonnet, who after years of experiment and devotion, including bankruptcy, finally produced late in the eighties a practical commercial fiber.

Chardonnet's process used cotton linters (the hair of seeds) and cotton waste from combing machines, and reduced it in alcohol and ether, finally changing over to the use of sulphuric and nitric acids and denitrating the yarns when wound on the cones and bobbins. This fiber was first shown to the public in the Paris Fair of 1889, and from here begins the real commercial history of the fiber.

In 1892 three Englishmen, Cross, Bevan and Beadle, developed a cheaper process through substituting woodpulp for cotton, and this process, generally known by the name of "Viscose," occupied for many years the principal place in the quantity market.

These have been the two basic methods upon which all subsequent processes have been developed.

But the demands in later years for quality fiber have swung the pendulum of trade sharply in favor of the general principles developed by Chardonnet, and most of the artificial silks now attracting attention are developed from a cotton basis.

All over the world today, in Italy, Belgium, Germany, France, England and the United States, there are fifty-seven large concerns in this industry, and the more or less accurate statistics of 1922 give a total output of seventy million pounds. When it is considered that only sixty-three million pounds of cocoon silk were produced in the same year, the predominance of this industry in a generation will be appreciated.

In America the Viscose Company for a long time was the largest producer of artificial fiber and was devoted to the woodpulp basis. This company has recently changed to a cotton basis and produces excellent yarn by the new process.

The American Tubize Company, an offshoot of a Belgian parent, developed finer yarns of greater strength in comparison to diameter, and of greater strength when wet than the viscose.

After the War, the Dupont Company developed an organization for making fiber silk and the Industrial Fiber Co. followed suit.

Latterly, the chief disadvantages of artificial silks have been their great diameter as compared to the real silk, which gave to cloth a rather coarse texture, and the fact that under moisture they lost about seventy per cent. of their strength. This was a great disadvantage until the public became accustomed to the fiber and learned how to wash and clean it so as not to impair its value.

Lately an English process, known as acetate silk, and called by the trade name of "Celanese," has attracted considerable attention. This silk is absolutely waterproof and does not lose any strength, therefore, when wet. The chief difficulty lies in the fact that it is difficult to dye, since it is impervious to water; but this disability has been largely overcome by inspired chemical research.

All of these fibers have peculiar merits and uses, and their great development during the last few years is proof sufficient that the public has accepted them without question. One of their great advantages for the weaver lies in the fact that they can be cross-dyed with cottons, silks, or wools, or in combination with any or all of these fibers. That is, if the proper fiber for the purpose is selected, a cotton or silk cloth can be woven, immersed in a dye bath that will dye the cotton or silk, but leave the artificial fiber undyed, and then immersed in a second bath which will color the fiber but leave the cotton or silk unstained. Very often it is possible to combine the chemicals in the same bath so that two colors are produced in a single dyeing.

Many changes can be run on this same theme, and it offers new opportunities for the designer and styler. Obviously, this is another intricate chemical problem and it must be admitted frankly that the average cotton or silk mill, as the average knitter and dyer, have not been particularly careful in studying out the peculiar qualities of each fiber and selecting the one best fitted for their needs.

A greater understanding, however, is growing between the fiber producer and the textile manufacturer and there is no question that artificial silk of different kinds will play a very large part in the future development of the style business in cotton fabrics.

In a general way, research in mechanical, physical and chemical fields has been accepted at least in theory, and the full practise will follow before long. Already in many ways American mills are the finest technical units in the world. More static labor populations will tend to vastly improve conditions. Such superiority as exists in certain phases of the British industry is due to longer concentration on some definite problem with greater specialization of machinery, and labor trained from generation to generation in specific tasks.

The common and perhaps natural habit of looking to a tariff to equalize these advantages, and a little too hasty tendency to blame differences in wages and hours of employment in Europe or other sections of this country for conditions have diverted attention from the essential exhaustive technical analysis and modification of existing systems.

William R. Basset, the well known industrial specialist, recently made a survey of the textile industries for the Hoover Committee on the Elimination of Waste in Industry. In time his keen analysis of conditions will receive a wider attention from the textile trade than it has so far perhaps attracted. He dwelt with special emphasis on the wasteful methods of buying raw materials, and on the evils of speculation in cotton. But his observations on the training of labor and a greater human coordination seemed to me to be of peculiar significance. Here indeed lies a great field for intensive study.

From time to time our cotton industry has been affected by needlessly bitter and rather aimless industrial feuds. These difficulties are in a measure due to the great diversity of textile labor populations and to a large percentage of absentee ownership. They are largely unnecessary and always very costly. It would seem as though some method short of a mere temporary balance of antagonistic forces might be found. There are, of course, large mill units as well as small, where labor difficulties have been reduced to a minimum. Such problems as face us today are of an economic rather than a human character. Few mill men today look upon labor with the cynical eyes of the elder generation. Equally with all minutely divided industries, the cotton mills are open to the attacks of disciplined and organized labor and might easily become the myriad theatres for tragic and sordid dramas of destruction. Saner councils will, however, prevail. The older, ultra-conservative die-hards and the uncompromising radicals will, I trust, never be permitted to make this industry their battle ground. If the progress of the last ten years is any guide, the cotton industry is oriented towards some rational form of industrial democracy.

With extreme reluctance I leave the discussion of these problems, which I have only suggested. They are of vital moment. Yet, since they have already been raised and are now under consideration in all sections of the country, I feel it wiser to devote the remainder of this chapter to a field of research only just beginning to attract serious attention.

If I have not suggested in this narrative that the history of cotton is the history of a series of great artistic achievements; if I have not proven that the basis of its very existence over long centuries of commerce and invention has been the history of beauty; then I have marshalled my facts with poor generalship.

The mere coincidence that the cotton fiber lends itself almost perfectly to mechanical production does not alter these facts in any way. The world did not tire of beauty because a few machines happened to be invented and perfected by one single people in the long histories of this great fiber.

The next chapter in the story of cotton will be written by the artist-designer. He is almost a new figure in this industry, but the scene is set and the audience waiting for his appearance.

Let us admit with candor that the great bulk of so-called textile designs not alone in cotton but in all fabrics, not alone in America but in England, have been mediocre in artistic merit for at least fifty years. There is no need to go over the dreary reasons for this condition. It is enough to write that the problems of conquering the machines, the development of world commerce, and the extremely delicate human adjustments to the new philosophy of automatic productive forces were gigantic tasks. A century and a half is in point of fact a very short time to allow for even their partial solution. With a few honorable exceptions in England and America, most of our designs have been obtained in one way or another from France, particularly Paris. In no sense do I mean to belittle the artistic achievements of a gallant and appreciative people when I write that the supremacy of France in the fields of decorative art is in a measure the result of her political exigencies during the Eighteenth and early Nineteenth centuries.

When the great industrial revolution began in England with Kay's fly-shuttle in 1733, France was at the brink of a series of national blunders and disasters which almost broke her spirit. When English manufacturers were perfecting machines and systems of mechanical production, France, bankrupt and ringed with foes, willed that kings should die. The year that Eli Whitney's genius gave to our South her great plantations of cotton, saw France, mad with terror, strike at her nearest foe. And while mills were springing up beside the turbulent rivers of New England, the steel tipped legions of France followed the brilliant Corsican victorious against a world in arms.

After the Napoleonic incident, followed the patient, bitter years of reconstruction, and it was not until after the Commune in 1850 that France began to take again her place in the industrial and commercial world.

The age of the machine had come, had grown in stature, and France had spent her energy in other channels. She faced a world gone mad over whirling machines and vast forces of productive energy. She had only her craft guilds, a tradition of honest workmanship, and a love of beauty as her salvation. She had never exchanged the artisan for the mill hand, never subordinated the artist to the mechanical director. Her small, personally supervised workshops filled with skilled craftsmen, proud of their dexterity, were more flexible mediums for artistic creation than the great mills of her sister democracies. The traditions of beauty that had grown up under the great Italian masters from the stormy times of François, the First, she had never lost even in her darkest hours of struggle. And so the world came to Paris in a stream of gold for the work of her masters. And France was wise, for no specious pleas from the advocates of serial production ever won her from her firm position. She kept her craftsmen and only experimented in a cautious manner with the machine. She gave security and honor and high rewards to her successful artists in all fields. She built schools and academies, workshops and museums to the end that this creative fertility might live and prosper.

England and America and indeed the entire industrial world fed their hungry machines with the ideas that first saw perfection in France. So long as France created, the machines could not become sterile for ideas. France, for almost a century, has been the world's studio.

If any valiant and patriotic manufacturer of fabrics in England or America doubts this statement, let him study the list of professional buyers of style merchandise who sail on almost every steamer to the ateliers of Paris, or look at the labels on merchandise in Bond Street shops. If any lay reader believes these statements too strong in regard to cotton fabrics, they have but to go into the nearest department store and ask to see the cottons of M. Paul Rodier, artistic heritor of the great craft ages and international merchant.

Then came the great War, and France gave every ounce of her energy, every atom of her power to divert disaster. Her craftsmen became Poilus, her masters gave their genius to munition making and the thousand and one grim concerns of war. At once the world's machines felt the interruption of the life-giving ideas. Manufacturers all over the world soon learned the lesson that merchandise is purchased on its artistic merit rather than its physical qualities.

In 1915 we felt this lack very keenly in America. E. W. Fairchild, publisher of *Women's Wear*, a daily paper with a national circulation to the retail stores and costume manufacturers in America, asked me if some means might not be developed here to fill the gap. There was no time to develop any complicated organization to build solidly from experimental beginnings. What was needed was swift action. Albert Blum, Treasurer of the United Piece Dye Works and a partner in a great Lyons dye house, was called into consultation. He at once agreed to the seriousness of the situation, and placed his time, energy and prestige at the disposal of any plan that remotely promised success.

Two great questions at once arose. Did America possess in its museums the collections of decorative arts essential as inspiration to designers? Was there in America adequate talent to reinterpret this material into acceptable designs?

The first question was easily answered. Our museum collections were adequate and accessible. More than this, the directors of our great museums in New York City and Brooklyn were more than anxious to assist us to any degree. In many instances they anticipated our requests and in all cases they had prepared long in advance of any sign of industrial interest. Our museums offered their collections, and still offer them to the industry with greater freedom than any museums in the world.

The next question was not so easily answered. For generations our designers had been discouraged. Textile pattern making was regarded as the lowest and worst paid of the arts. Quite properly our art schools almost totally ignored it since it offered absolutely no field for ambition. England was a little better, perhaps, but only to a degree. Some mills had small staffs of ill-paid copyists or hack designers. There were a few commercial studios that produced painted bits of paper about as inane and original as tomato-can designs. Was there, then, in spite of this colossal and wasteful indifference and neglect, in all this broad land, talent for our needs?

First we threw open the museum collections and published in *Women's Wear* and elsewhere information to the designers advising them of the great facilities before them. Next a little booklet was prepared showing the technical details of mechanical pattern-making, and finally a few hundred dollars set aside in prizes by *Women's Wear* and a nation-wide contest in design organized with the assistance of the art schools of America.

The first jury met in the Metropolitan Museum of Art in the early winter of 1916. I well remember the closing hours, with a blizzard raging outside, when the belated artists brought their bundles of designs to my laboratory in the American Museum of Natural History. I can still recall my anxiety, for until the last three days there was nothing to prove that all of our plans and preparations were not in vain. More than the mere success and failure of a plan was at stake. I had to reckon with the enthusiasm of men in a project from which they themselves had neither hope nor expectation of reward.

Dr. Clark Wissler of the American Museum of Natural History had said: "It is a source of wonder to me that our textile designers have not made freer use of the collections under my care. Certainly they should serve as a wonderful inspiration."

In this connection it may be said that lectures were given in the American Museum of Natural History and the Metropolitan Museum of Art by distinguished scholars on the problems of decorative art and by technical experts on the relation of the machine to design. To these lectures designers and design students had been freely invited. There had been as well considerable missionary work in the art schools, both along the line of research in design and the practical application of design through mechanical methods to fabrics.

Henry W. Kent, Secretary of the Metropolitan Museum of Art, said: "We have always been anxious to reach the great class of textile designers. Individuals have used our collections for a long time, but as a class designers have held aloof. The attitude of *Women's Wear* in this matter is admirable, and the effort to give expert advice should meet with success."

Albert Blum said: "It is time that America developed a distinctive textile art. The opportunity to study collections in the two museums through these lectures is wonderful."

The first jury consisted of Henry W. Kent, Albert Blum and the author. A few hundred designs were submitted for our inspection and of these about one hundred and seventy-five were hung up in an exhibition to which the textile manufacturers were invited. Even in this first trial the American designer acquitted himself well; or I should have said *herself*, since most of the prizes went to women. Both professionals and amateurs had an equal standing before this jury, but the amateurs won most of the prizes then as later.

Five such contests in all were held, the last four under the auspices of the Art Alliance of America, who generously donated their rooms for the annual showing and took care of the details of the contests. Beginning with the second contest, certain individuals in the industry made generous contributions, and many

RESEARCH

1—Detail of one of the Design Research Rooms in the Brooklyn Institute Museum. (*Page 204*)

2—Exhibition of "Cottons Ancient and Modern," in Mechanics' Hall, Boston, National Association of Cotton Manufacturers. November 1st–6th, 1923. (*Page 205*)

3—Fifth Women's Wear Design Contest poster, November 8th, 1920. (*Page 197*)

4—Page from Women's Wear of July 23rd, 1921, illustrating the possibility of Oriental arts in cotton. (*Page 199*)

5—Page from Women's Wear of July 23rd, 1921, illustrating the connection between India, New England and the South in cotton designing. (*Page 199*)

6—Page from Women's Wear April 27th, 1922, illustrating the value of the Decorative Arts of the South Sea Islands as a source of inspiration for cotton designing. (*Page 199*)

7—Page from Women's Wear, July 2nd, 1921, showing ancient documents and modern adaptations. (*Page 199*)

PLATE 21

expert judges of design either served on the jury or acted in an advisory capacity. The last of these exhibitions was held in 1920, and over one thousand artists, representing thirty-four states and Canada, sent in thirty-five hundred designs, and received prizes of over twenty-three hundred dollars.

In all of these contests the designs remained the property of the artists and since all prize designs as well as many others always sold, the successful designers received substantial rewards, and often a sustained recognition in the industry. A list of the jurors and contributors and the consistently successful artists follows:

Jurors and Contributors

HENRY W. KENT, Secretary, Metropolitan Museum of Art
ALBERT W. BLUM, United Piece Dye Works
EDWARD L. MAYER, Costumer
ALBERT HERTER, Herter Looms
J. H. THOMPSON, B. Altman & Co.
W. G. BURT, Marshall Field & Co.
MILTON VOGEL, Bonwit Teller & Co.
E. IRVING HANSON, H. R. Mallinson & Co.
CHARLES GOWING, Burton Bros. & Co.
J. A. MIGEL, J. A. Migel, Inc.
CHARLES PRENDERGAST, Artist
FREDERICK C. FOLSOM, F. A. Foster & Co.
MAX MEYER, A. Beller & Co.
STEWART CULIN, Brooklyn Museum
J. W. MACLAREN, Johnson Cowdin & Co.
HARRY WEARNE, Interior Decorator
HENRY P. DAVIDSON, Interior Decorator
GEORGE B. CHADWICK, Associate Editor of Country Life
HARRY NEYLAND, Swain Free School of Design
F. W. PURDY, Art Alliance of America

CARL ROESSEL, Louis Roessel & Co.
CHARLES CHENEY, Cheney Bros.
F. W. BUDD, Cheney Bros.
M. D. C. CRAWFORD, Associate in Textile Research, American Museum of Natural History

Artists

MLLE. DURANT DE SUMENE
MARTHA RYTHER
ADRIEN FLEARY
MARIA C. CARR
ALICE M. HURD
MARGURITE ZORACH
FRANCIS FULTON
HAZEL BURNHAM SLAUGHTER
GRACE H. SIMONSON
BERTHA MOREY
LILLIAN LAWRENCE
HELEN S. DALY
LAURA E. WALTON
BERTHA SMITH
RUTH J. WILSON
W. E. HENTSCHEL
ZOLTON HECHT
MARY TANNAHILL
NELL WITTERS
FLORENCE LONG
RACHEL SMITH
ANNA PIKULA
ALPHONSE BIHR
HOPE GLADDING
PIETER MIJER
ALICE F. TILDEN
MARY MARSHAL
JULY CONE
MARY LOUISE CLENDENNING

EMMA W. DOUGHTY
KATHERINE W. BALL
BESS B. HUGHS
LOUISE DREW
Y. CONSTANCE DUFFY
EDNA B. LOWD
CHOLLY FRIETSCH
MARION POOR
CLARICE PETREMONT
COULTON WAUGH
ILONKA KARASZ
RUTH REEVES
F. WEINOLD REISS
A. J. HEINKE
CONRAD KRAMER
FANNIE BAUMGARTEN
ALICE L. DALLIMORE

Very early in these contests it was discovered that many talented designers had greater facility in working out their ideas on actual fabrics than on paper. A supplementary competition was therefore, arranged, known as "The Albert Blum Exhibition of Hand Decorated Fabrics," and many of the most beautiful designs were developed in this way. Hand craftsmanship was found to be of the greatest importance in the problem of design, and this we should have known from the history of ornament as well as from practical experience.

I have written before of the great industrial service rendered by the museums. But the Brooklyn Institute Museum played so vital a part in the later and more intimate developments in American textile and costume designing that some special mention seems appropriate. With the generous support of the trustees and working in cooperation with the leaders in the industries, Stewart Culin, Curator of the Museum, organized the most complete research collections in the world for the purpose.

Certain rooms and special collections of materials and books were set aside for practising designers. Many of the most gifted creators in all fields make constant use of this material, and even foreign designers have come to know of the opportunity and gladly and freely take advantage of it. Everywhere through the industries, I can trace the direct or indirect effects of this inspiration. No collections in the world are so accessible, no museum has ever made fewer conditions for the uses of its material; nor has the full usefulness of this work been as yet appreciated. Every year it grows in extent and value and no one may write at a future time of the American Renaissance without full recognition of the work of this museum. It is beyond praise, as it is above reward.

Very soon the use of historic materials began to spread beyond the industries into broader public fields. In the fall of 1921 I was asked to prepare an exhibition of the history of art in cotton for the Cotton Machinery Exposition in Greenville, South Carolina. With the aid of the Brooklyn Museum, the American Museum of Natural History and my own private collection, an exhibition of the history of art in cotton was organized, put in charge of a competent assistant and sent to Greenville. The retail stores learned of this Exhibition and immediately from all over America we received requests for it. This was embarrassing. The material was very precious and to a degree fragile, but arrangements were made and the collections visited many cities. There have been four exhibitions of cotton alone, counting the first one. The last one was organized at the request of the National Association of Cotton Manufacturers in November, 1923, and was shown at an Exhibition in Mechanics' Hall in Boston. At this writing this exhibition is still on the road, having been shown in the following cities:[1]

BONWIT TELLER & CO.	New York, N. Y.
NEIMAN-MARCUS CO.	Dallas, Texas
GUS BLAS CO.	Little Rock, Ark.
L. R. EAKIN	Manhattan, Kan.
STIX, BAER & FULLER CO.	St. Louis, Mo.
KAUFMANN STRAUSS CO.	Louisville, Ky.
THALHEIMER BROS.	Richmond, Va.
FRANK R. JELLEFF, INC.	Washington, D. C.
HOCHSCHILD KOHN & CO.	Baltimore, Md.
KAUFMAN & BAER CO.	Pittsburgh, Pa.
C. F. HOVEY CO.	Boston, Mass.
THE HOWLAND DRY GOODS CO.	Bridgeport, Conn.

There were forty requests in all, and this exhibition is still to appear in the following cities:

LA SALLE & KOCH CO.	Toledo, Ohio
THE MOREHOUSE-MARTENS CO.	Columbus, Ohio
THE RIKE KUMLER CO.	Dayton, Ohio
H. & S. POGUE CO.	Cincinnati, Ohio
L. S. AYRES & CO.	Indianapolis, Ind.

[1] Previous to the exhibition of the National Association of Cotton Manufacturers, a collection of cotton and cotton dolls known as "Thirty-Nine Centuries of Cotton Development" was organized." This was first shown in the retail store of Carson Pirie Scott & Co. in Chicago, the weeks of January 29th and February 5th, 1923, and was presented with great distinction by this organization. Thousands of people came during the period of two weeks to see this collection and this was the first time that a complete story of the history of a single fiber had been presented to the public in connection with modern merchandise.

WHITNEY-MACGREGOR CO.	Minneapolis, Minn.
EMPORIUM MERCANTILE CO.	St. Paul, Minn.
FREDERICK & NELSON	Seattle, Washington
MEIER & FRANK CO.	Portland, Oregon
THE A. T. LEWIS & SON DRY GOODS CO.	Denver, Colorado
MILLER & PAINE	Lincoln, Nebraska
DAVIDSON BROS.	Sioux City, Ia.
THE DENECKE CO.	Cedar Rapids, Ia.
YETTERS	Iowa City, Ia.
HARNED & VAN MAUR	Davenport, Ia.
L. H. FIELD CO.	Jackson, Mich.

There were one hundred and sixty-seven requests for the exhibitions of cotton materials, but all of these could not be satisfied. The time that such specimens might be away from the museums was, naturally, limited. These exhibitions have been shown, however, in forty-eight cities all over America, and in many of our great educational institutions. In this way materials that formerly were seldom seen outside of great metropolitan areas have had a wide circulation throughout the entire country.

These exhibitions have unquestionably awakened the public to the great possibilities of the cotton fabric as a medium of artistic expression. They have given to the mills some indication of the possibilities of design, and have served in a degree as an interpretive medium between the public and the mills.

It is impossible in this brief space to enter into all the ramifications of this movement. Many wholly worthy phases of it I have left untouched. In the main, what has been accomplished so far has been to call attention to the museum collections as a source of inspiration and to establish the American designer on a somewhat more secure footing, and to bring to the cotton mills a realization of the part that design might play in their success.

With these facts in mind, and with the record of experiences successful and otherwise behind me, it is possible to outline, in broad terms, the next development.

It is plainly evidenced that there never can be any serious advance in our decorative arts without adequate and accessible historic material as a basis of inspiration. In writing this I do not wish to stifle originality in any degree. To merely slavishly copy old designs is not enough. There must be creative interpretation. Each designer, according to his or her imaginative force, training and appreciation, will absorb from the ancient arts ideas and give these expression in new beauty. It has always been so. The history of all art, upon the surface at least, has been the result of exterior impulses resulting in fresh creative power. The ancient arts of Asia Minor and Crête aroused the Greeks; Italy learned from the Near East and Greece; France and Spain drew inspiration from Italy and the Moors; England and America followed France. Forms of expression are nationalistic only so long as they are in the process of creative evolution.

Back of all these historic art contacts lie the great periods of expression whose vague antecedents we can trace in ancient wall paintings of Ajanta in the rich treasure tombs of Egypt, in the sandy graves of Gobi. Always some past beauty has been the teacher and inspiration of new loveliness.

The ideal designer of the future must then be something more than a student of art history. Trained draftsmanship is essential. And beyond this must be an understanding of the potential capacity of the machine as a medium of expression, and a sympathetic accord with the technician. When these desiderata have been accomplished, there is still to be mastered the difficult problem of gauging the public taste which we loosely refer to as style.

The whims of peoples for any particular kind or type of ornament, the preference for certain colors, or texture are never wholly accidental, nor to be lightly regarded. This stuff of which dreams are made is in sober truth the very essential of all culture. Through long ages men have desired beauty, have striven to express their ideas of charm. If this desire ever left us for an hour, that hour would mark the wreck of civilization.

So the historian of art, the draftsman, technician and style expert must be combined in the designers of tomorrow. Short of this we will never achieve our destiny and the machine will continue to lose prestige and scope.

Every texture, every type of design, the complete palette of color, the highest standards in workmanship belong not to the machine but to the craft ages. It is, of course, in one sense unfair to bring the expression of a single age in comparison with the accumulated arts of many ages and many peoples.

The great disparity between the æsthetic values of machine-made materials and the priceless documents of many yesterdays is due to the fact that in craft ages there was the opportunity for experiment and comparison, and a closer relationship between production and creation than in our own times. There are for us new constructions of fabrics, new ideas in technical arrangement of yarns and patterns, that are still to be worked out on the machine from research. So far as surface design is concerned, particularly as this applies to the printers art, the records of art history offer an endless inspiration. There is no excuse except ignorance and indifference for any unlovely printed fabric, regardless of its price or quality.

I believe, however, that the most important subject is that of color. It is so easy to confuse in discussion or writing, the names of colors with the sensation of colors themselves, that the vital importance of the subject is seldom understood.

Each season the public focuses its preference on some rather limited range of shades, and the great bulk of buying is always on a very narrow section of the palette. The fact that this preference is well understood, and is indeed carefully fostered through the few avenues of publicity used by the textile manufacturers, has never apparently induced any individual or group to make an exhaustive study of the reasons governing this preference.

Any careful analysis of the situation will show that the colors that reach and hold public favor are almost always those that have been by some agency developed from a traditional source. Good color is never an accident. The reaction to color is one of the most vital and one of the most ancient of our emotions. It is quite evident to all students of human culture as to all style experts that a delight or the reverse in chromatic effects is the dominant impulse in our relation to decorative arts past or present.

Any museum is in one sense little more than a glorified laboratory of color. It can not be used intelligently except by trained individuals, but the material for research is always present. In spite of these facts, annually the mills of this country and England dye immense quantities of merchandise without any particular attention to the quality of color except such as are familiar to the chemist.

Our public has become highly critical in regard to color values and will no longer accept shades or tones unless these are distinctive in charm. It is, fortunately, no longer possible for mills to violate the traditional canons of good taste since these now run counter to the social instincts.

I am conscious that I have glorified an ill paid trade into a profession. This was deliberate. The designer is the artist working through the medium of the machine. Either this or nothing.

The problem is not, do we need such a profession; but rather can such individuals ever fully express themselves directly through our intricate machinery? Do we not need some intermediate agency, some kind of design laboratory for first experiments? The mere fact that in England and America we have great mills and precise mechanical rules governing production does not mean that we may not develop craft shops as well. As a matter of fact, the hand craftsman is on the increase in both countries and will grow with developing skill and broader understanding of his opportunities.

Annually we import millions of dollars worth of hand craft fabrics, and our own craftsmen are by no means denied their markets. Each great unit of mills needs its own special craft laboratory to make materials, to adjust them to markets by tentative trial, and to arrange them for the machine after they have won acceptance.

Hand craft is not only a more certain, but is the cheaper method in developing new ideas in textures and patterns. The machine only becomes economical when large quantities of materials are wanted. New ideas seldom win immediate public approval. They must first win the acceptance of a rather small and highly critical group of patrons. Here the craftsman is supreme.

It is essential that the craftsman be accorded full legal protection for his creative effort. Our indifference to the rights of designers amounts almost to a national scandal. Copying successful patterns is the meanest form of commercial dishonesty, and for the entire industry the most expensive, since it discourages the creative force which in the last analysis is the vital life of business as of art. A law giving ample and reasonable protection is one of our great needs. In a nation addicted as we are to law-regulations, it seems surprising that no such statute has been enacted. Every time some form of protection has been suggested it has met with opposition on the part of certain elements, and been supported in a luke-warm measure by others. This is, of course, a kind of left-handed recognition of the immense importance of design, if a very poor method of encouraging design.

A law that will yield protection and at the same time place no unnecessary restrictions on the use of historic materials is needed. It will not be an easy law to draft, nor can it be safely left to our professional legislators. A committee of experts should be organized to study this problem in all its different phases and a broad public spirit aroused in its enforcement. Every time a design is copied and cheapened, both the designer and the public are defrauded.

These plans are by no means simple. I never intended to convey any idea that the correction of ancient evils could be accomplished by merely wishing. But they can be corrected. We can have beauty and charm in cotton materials produced by machines at less cost of energy and money than is now wasted through errors in artistic judgment, in machinery, or in attempting to seek protection of profits in other directions.

I do not feel it to be wholly a matter of choice, nor do I appeal to the spirit of public service in cotton mill owners. It is my sober belief that in the intelligent solution of artistic problems lies their only salvation. This world has not been cured of its love of beauty because some few men own cotton mills.

CHAPTER XV

CONCLUSION

IN the introduction to this volume, I promised to keep the narrative as free as possible from technical details and statistics. There are, however, certain facts which can be presented through no other medium.

The United States still far outdistances the rest of the world in the gross weight of cotton raised each year. In 1922 with the world cotton crop around seventeen million bales of five hundred pounds each, we produced over ten million bales. Egypt, India, China and Brazil followed in order. These figures, as all broad generalities on staple products, are open to distinct view-points. As a rule the mill buyers believe government estimates too low and farmers and merchants believe them to be too high. The total figures never include the many hundreds of thousands of bales consumed in the domestic markets of China and India. The year 1922 is generally regarded by experts as a very short crop, several million bales below the highest world average such as 1907.

In regard to the quality of the fiber as determined by its length, fineness of diameter, spinning character, color and freedom from foreign matter, James McDowell, one of the world's leading authorities on cotton, has furnished me with the following data:

The finest grade of cotton is grown on the islands off the coast of the Carolinas and Georgia. This crop was at one time of the greatest importance in the spinning of extra fine yarns and sewing threads. Due to the ravages of the boll weevil, this crop is very small today, seldom exceeding two thousand bales. Next in quality is the Sea Island Cotton, grown on the Islands of St. Kitts and the Barbadoes, most of which is shipped to England and France for the lace trade. The great fine staple of commerce, the fiber most often used where mercerizing of yarn and fabric follows, is that grown in Egypt. After this follows the long staple American cotton grown from specially selected seeds in favored farming regions on the coast of Florida and the rich lands of the Mississippi Valley and certain parts of Arkansas and Texas. The Soudanese, Brazilian and the Peruvian, both rough and smooth, follow. After these come the great mass of the American short staple, upland cotton, the Indian and the Chinese. Cotton cultivation is being encouraged in Samaria in West Africa, in Russian Turkestan (for the Soviet Government) and in Turkey. Under subsidies from the British Government, cotton is being cultivated on the Gold Coast of Africa, Lagos, South Nigeria, North Nigeria, West Africa, British Uganda, British East Africa, Nyasland, Rhodesia, East Africa and Soudan. The most important experiment now under way is, however, in Western Australia and Queensland, where under careful governmental restrictions, Chinese coolie labor is being employed. The yield from these farms is about eight hundred pounds of lint per acre as compared to an average in America of under two hundred pounds per acre. The high prices prevailing for cotton during the last few years, have also en-

Conclusion

couraged cotton growing in Southern Mexico, Yucatan and Ecuador.

When it is realized that from a non-producing cotton country in the late Eighteenth Century, by the middle of the Nineteenth Century, cottons of America dominated the world, it will be realized that our supremacy as a cotton raising country may easily be challenged in any generation by any of the continental areas I have mentioned above. Egypt's supremacy, on the other hand, in fine qualities of cotton is by no means assured. The splendid control of the British Government for many years directed the seed selection and cultivation in Egypt; and if the present unrest in Egypt destroys this control, it will be very difficult if not impossible to restore it, except after long and very expensive experiments. So far no field labor in the world compares to that of the negroes in the South, and this is really the dominant factor in maintaining our control of the fiber. Latterly the shortage of negro labor during the picking season has led to the introduction of casual labor from Mexico, and while this labor has been in the main satisfactory, it does not compare to the trained negro cotton picker.

There is no menace, however, to our cotton industries suggested in the possible loss of our supremacy as a cotton raising nation. England, the mother of the modern cotton industry, still maintains her vast superiority as a producing center and has never raised a single pound of cotton and never could, outside of a green house. The United States is today by far the greatest producer of silk goods, and all her raw fiber comes from Japan, China, India and Italy. The cost of transportation of cotton in the bale is very slight compared to the value of the finished product, and manufacturing supremacy will be determined by the trained technicians, market conditions, intelligent merchandising, skillful and satisfied labor and designers. No one familiar with the average cotton farm and the attitude of mind of the average cotton farmer can be surprised at the discontent with cotton as a staple crop. It is the last of the great staples to resist the use of labor saving agricultural devices. It is the most unintelligently marketed crop of all the great agricultural commodities. It requires longer and harder work and there is less security as to rewards than any other. The general condition of the southern farmer so far as life and luxury are concerned is bad.

Out of the fertile, inventive genius of America may come in time a practical cotton picker. Experiments with machines of this kind are already as far advanced as was the automobile fifteen years ago. There is then to be considered the control of the boll weevil, the little insect which annually destroys over three hundred million dollars worth of cotton lint. A larger percentage of graduates of agricultural schools actually on the land and the use of better methods of cultivation and more suitable types of seeds will help the situation.

But the average farmer is far from convinced that any measures tending to increase the yield per acre or increase acreage under cultivation are benefits. A bitter experience has proved to him that the large crop, so ardently desired by manufacturers and brokers and so desirable perhaps from a general, economic viewpoint, means for him low prices and little profits, and often a crushing burden of personal debts. In many parts of the cotton belt, the boll weevil, looked upon by scientists and the public at large as the enemy of the farmer is regarded by the farmer as his friend, and he

Conclusion

has no real sympathy with any of the many plans now advocated to destroy the pest. It is easy enough to meet his prejudices with theories, but it is hard to arrange arguments to answer the actual facts of his life.

There has been a great improvement in warehousing for cotton, more intelligent banking methods are in vogue, there is the beginning of scientific experiments in seeds, and a closer sympathy between the trained mill buyer of cotton and the farmer and all of these things I reckon in the list of benefits. It is now possible for the farmer to borrow money outside the limits of his own community on receipts from bonded warehouses, and this has freed him from a system of money lending that would make a pawn broker ashamed. And all of these things are excellent and fruitful of much of the recent prosperity in the South. The cotton farmer looks, however, with a brooding suspicion on the world's cotton exchanges and in many instances this suspicion is justified.

The buying of the actual cotton fiber is beginning to be done with some pretense of scientific analysis, and of course the natural and healthy conflict between buyer and seller is constructive and helpful. But on the exchanges of the world, future contracts for cotton are annually made totaling unknown hundreds of millions of bales, while the actual cotton crop of the world, as I have mentioned above, averages between sixteen and twenty millions of bales. This anomaly is difficult to explain to the lay mind and, as a matter of fact, to a large extent it has no valid explanation. In the main it is a vicious form of gambling. A certain percentage of this speculative buying is of course a kind of insurance policy taken by the seller or manufacturer of cotton goods against future orders, to protect himself against the mercurial fluctuations in the raw market, and this activity on the exchanges is obviously legitimate.

The principal business, however, of the cotton exchanges is purely speculative and highly dangerous and has no more to do with industry and beneficial economic conditions than gambling at Monte Carlo or the race tracks. Cotton exchanges are fed by a net work of telegraph and cable wires reaching not only to the financial and industrial centers, but to each little farming village. The mania to gamble in cotton futures is just as strong among the Arabs along the Nile as the cropper farmers of the Mississippi Valley or the Texas Plains. Since speculators of this class are usually (for some unknown reason) optimists, the fall of cotton prices of a few points means a heavy toll in human misery. In its worst form gambling in cotton futures is the meanest kind of bucket shop operation. It neither aids the farmer, the mill man nor the public. It is purely parasitic, even when legal. That men should work and drive their wives and children to the verge of desperation and then lose their meager earnings by any such means, can have no serious apologists.

Gambling in cotton futures is not confined, however, to the unwary agriculturist. It is a common habit unfortunately with mill treasurers, who should know better, and the experts in the broader ranges of speculative finance, from time to time, are impelled, in spite of previous experience, to test the quickness of their eye against the gyrations of this elusive pea.

I am in perfect sympathy with the sincere men who see in better seed selection better warehousing, banking and growing, in the regulation of exchanges, and the elimination of cotton gambling, concrete benefit. But

Conclusion

back of all of these questions, lies the still broader problem of land ownership. The vast majority of cotton is grown on rented farms, the lessee or the cropper giving from one third to one half of his crop as rent. He is furnished with seeds, at times farm implements, under certain conditions perhaps the use of a mule, some tragic apology for a home, and he is financed at ruinous rates during the planting and cultivating season. He is a victim of the vicissitudes of the market. Seldom can he hold his cotton for a favorable price, for by the time it is picked, ginned and baled, his debts at most ordinary seasons have eaten up his entire equity. He is prevented, by the owner of the land in most instances, from growing anything else but cotton, and is compelled to buy from the local store the food which he might more profitably and easily raise himself. I know there are exceptions to this rule; I know of men of vision and humanity who handle their estates with constructive judgment, but the general condition of the cotton farmer in the great cotton sections of America is little if any improvement above chattel slavery.

The greatest improvement that could come to the prosperity of the South, the greatest insurance this nation could take out to protect its supremacy as a cotton raising country would be a change in the system of land ownership in the South by intelligent banking facilities, which would permit the farmer to own and control the land on which he works. This in my judgment is the key to the situation. There is nothing so good for farming as having the farms owned by the men who work the soil and from such a condition all the other benefits now so ardently advocated by the friends of the farmers would easily and naturally result.

220   𝔗𝔥𝔢 𝔥𝔢𝔯𝔦𝔱𝔞𝔤𝔢 𝔬𝔣 ℭ𝔬𝔱𝔱𝔬𝔫

In the year 1919 there was held the First International Cotton Conference. The world was then cotton hungry. The great demands of the wartime had exhausted the world's supply and a cotton famine was imminent. So the mill men, technicians, statisticians, bankers and technical experts from all the countries of the world, with the exception of Germany and Austria, planned a trip to the cotton centers of the South. This country, for the first time perhaps since the Civil War, was experiencing the stimulation of a sudden influx of wealth. Sleepy little villages had been converted into thriving prosperous towns, the streets lined with shiny, new automobiles, the stores full of high priced merchandise, hotels and theaters thronged with gay, extravagant crowds. In one little, dusty, Georgia village, a County Fair was in progress and cotton farmers, who until this time had never known any surcease from the burden of debts nor life beyond the limits of stark necessities, waited patiently in line to ride in an aeroplane at the rate of a dollar a minute. Everywhere was the suppressed excitement of an oil boom or a gold rush. Fortunes were made over night by individuals who had never known prosperity before. There was building of roads, of homes, of schools and churches. Field hands, who in ordinary seasons had earned perhaps sixty or seventy cents a day, were now demanding and receiving from seven to ten dollars a day. Patent leather shoes, silk striped shirts, phonographs and every other formerly forbidden luxury were being purchased in immense quantities. At the levee in Memphis, I saw a great cotton boat tied up at the wharf, the crew on strike for $7.50 a day and meals. We passed through miles and miles of cotton fields with cotton worth fifty to seventy cents a pound, rotting on

Conclusion

the plant for lack of pickers. In clubs and hotels, in banks and warehouses, men of affairs expressed the sober conviction that the day had arrived at last when the South had come into her empire and had no further need for New York, London or Chicago to finance her staple crops or industries.

In the little mill towns there was the same evidence of prosperity. Most of the three hundred and fifty visitors, who lived in pullman trains during this trip, were mill men, experienced mill men from the cotton factories of the world. They knew of cotton factories in the South, but had considered them as doubtful ventures, running intermittently, badly equipped and poorly managed. What they saw changed their minds radically.

Mills were running at full speed at high profits and high wages. We had all heard stories of oppressed and dissatisfied, underpaid and under-nourished labor in southern mills. We saw little villages with pretty cottages, gardens, town halls, athletic fields and nurseries and throngs of well conditioned prosperous workmen. There was in many of these towns to be sure, the evidence of newness. They were the product of the last few years of high prices and hungry markets. Everywhere was a spectacle of a people long denied the simple comforts, for the first time indulging their natural desire for luxury. Cotton had kept them poor, cotton had made them rich! The pent up desires of half a century of enforced self denial was seeking satisfaction!

We met finally in convention in the charming old-world city of New Orleans, one of the greatest cotton ports of the world. Here mill men, bankers, statisticians, technical experts and merchants met with the growers of cotton. No industrial convention I have ever attended had one tithe of the human interest of this one. Here was some lord and master of a million whirling spindles, well dressed, suave, alert, and there a sun-dried farmer, who tilled his acres in some Louisiana bottom farm. Here was a man from the well ordered life of middle England, who had spent a half a century in a cotton mill and had seen on this trip, for the first time, a field of cotton, and next to him a lean faced, keen eyed cotton banker from a Texas city, judging the world from his outlook over the dusty spaces of his native state.

Experts talked statistics, grading, finance, shipment, packing, warehousing, loom hours, better seeds, the world's presumptive needs for cotton for a generation. To each problem these men had a definite, academic solution, not untinged perhaps with self-interest. If the farmer would plant more acres, employ better methods of cultivation, exterminate the boll weevil, use better seeds, and raise ten times the crop he raised at present, they proved to him conclusively that his income would be greater, his prosperity more sustained, for they could sell his surplus product in the hungry markets of the world.

To these theories and platitudes the sun-dried men, who were giving their youth and the youth and hopes and the lives of their children and women folk to raise cotton, made answer. Let the world starve for cotton; let the mills stand idle for cotton; unless the world were willing to pay the price that cotton was worth. For if it lay in manhood, cotton must first yield to them the good things, the simple, good things of life, so ardently desired, so long delayed, so briefly enjoyed. Talk to them about laws of supply and demand! Go hoe a row of cotton under a blistering sun, and see your little

Conclusion

tender children working beside you in Tophet. Carry the grinding load of a farm mortgage through a few bad years, and see your children leave you and go to the city, or wither in want on the farm. Stagger under that weary, honorless load on the same road they had groped along and then talk to them about economic theories! Kill the boll weevil? He was their one friend, he gave them schools and clothes and food. Large crops meant wealth to the merchants of cotton, to the mill men, cheaper clothes perhaps to the world, but to them it meant lack of the few creature comforts that life held for them. They met together, coatless, angry, unimpressed by all the waiting world outside, not without force of rough eloquence, and swore to curtail, to drastically curtail their acreage.

This was another phase in the cotton story and one deserving of the most earnest consideration.

Many of the mill men, to their credit be it written, especially among the English, were not lacking in sympathy. They too had come up a long, hard road and painfully they knew at first hand long hours of toil and the endless round of the hopeless days. The fine traditions of the open forum, that has kept their little Island secure since Norman William conquered it, made them respect these hard, controversial knocks. They shook their stubborn heads. It was a great question and the answer did not lie, could never lie, they knew, in simply passing the burden of cotton on to these coatless, collarless men, burning with a sense of their accumulated wrongs.

I shall not add to the burden of these men the affront of some obvious, academic cure for all these ills. I am not entirely convinced that there is any sure solution within our power to apply. This is, however, certain, we should allow no undue tenderness for middle men or produce gamblers to stand between justice and their cause. It is easy to sit in some great center of population and sentimentalize on the farm question and perhaps our great wheat farms are in little better condition. If more education is needed, if better seeds will help, if improved machinery will solve some problems, or better warehousing and financing ameliorate conditions, in the name of decency let these things be given. If there is no answer to cotton but the continued misery of an industrious and sober farm population, then the sooner we cease to be the greatest cotton producing nation in the world, the better off we shall become.

In this as in other phases of cotton, I am an optimist. The problems are already in solution. We will find mechanics and research in agriculture and more intelligent merchandising and more humane systems of land tenure, more economical in the long run than human misery. Once such conditions are understood in a democracy they must be rectified.

The price of a single battleship, or a presidential election wisely spent in constructive investigation, might solve this problem.

I have touched very lightly on the question of labor in this narrative. This is from no lack of appreciation of its vital importance, nor because of any belief that all is well in this respect. As a matter of fact, there is no more important problem than that of the human relationship towards the mechanical organizations producing textiles. I would not have any words of mine distorted into a lack of sympathy with the efforts of the men and women in the labor organizations, who have worked earnestly and under conditions often of great

Conclusion

personal danger to ameliorate the distressing conditions, which existed in this industry within the last decade. I am equally solicitous to support the sincere and far sighted mill executives, who have tried to bring about some common meeting ground between conflicting forces. There is still much work to be done and many difficulties to overcome, and both types of progressives need all possible encouragement and sympathy.

In a general way, however, the economic problem is in a fair way of solution. The attitude towards immigration so strongly held in all parts of America, since the Great War, will make it impossible within this generation at least, to dilute our labor population with aliens with lower scales of living values. This will do away with that competition for employment, which formerly placed labor and the humane and intelligent employer alike at the mercy of labor mongers. A higher economic value will be placed on the human factor, and the energy and such genius as exists in superintendence will be directed towards better types of machines and processes to make labor more efficient and more productive. The immense sums formerly spent in proselyting alien labor, the contesting of strikes and lockouts, loss due to curtailed production might be spent to greater advantage in training labor to a better understanding of the machines and a higher sense of obligation to the public.

The South, from its ingrained social habits and the close relationship between the mills and the mill communities, will not, for the present, lift its ban against European colonists. Any attempt to reproduce in mill centers in the South conditions which existed in Lawrence, Massachusetts, and other New England mill towns of recent memory will be met with a fierce opposition, in which the entire community will surely join.

So both of our great textile sections will tend to static labor conditions and the pitiful traffic in ignorance will diminish, if not cease entirely.

If unhappily I am too optimistic in this summary, then we are laying the scenes for a social upheaval in which all industries will be affected. We were tending towards this condition when the War came, and were saved, through the inflation of values, from a predicament into which selfishness and stupidity had led us. In the same way I believe that the present differences in wages and hours of employment between the South and East will tend towards a common level and this tendency will be upward and not downward.

The record of textile labor from the machine age onward has been bad, distressingly bad. In New England, up to the great strike in 1912, which won a wide public sympathy to the workers, no class of American labor was worse treated. I do not know of any actual condition as terrible as the complacent, mechanistic attitude of the steel industry which demanded for manufacturing purposes that large groups of men should work twelve hours a day, seven days a week. But one thing is certain, the light physical character of textile work, the small element of personal danger involved made it, for a time, a fruitful field for the exploitation of women and children in industry. The later development in southern mills, so far as child labor was concerned, was at least as bad, probably worse than in the East. England, first in the field of mechanical production, created conditions of life in the midland counties, which Englishmen of conservative attitudes of mind have described in terms more violent than have ever been employed by the most desperate radicals in describing modern conditions. So it has not been a matter of a ruling race dominating weaker peoples who chanced to be within their borders. The tyranny of Englishmen over Englishmen, and of the South over its own native labor left nothing to be desired in the way of human misery compared even to New England's attitude towards Greeks, and Slavs, Italians and other races, who came to supply the human element in her mills. These conditions are still so recent that it does no harm to call attention to them and to encourage each man and woman who reads these words to resolve that in America such conditions never may be brought about again.

But even with the most ideal economic justice that may be gotten under our present system, or indeed under the most radical ideas that have been advanced and which are in a measure in practice in Russia today we have still a human problem in textile labor that must be met. The machine age is itself so recent an intrusion in culture that humanity has not yet become adjusted to it. It was never intended in the scheme of things that men and women, descendants of ten thousand years of craftsmanship should perform hour after hour, day after day, year after year, incomprehensible tasks over which they have no control and which offer to them no stimulus of understanding and sympathy. Anyone of sensitive nature, familiar with textile manufacture, must have a strong moral sympathy for the great idealistic revolution of John Ruskin and William Morris. These philosophers saw clearly that the problem was not entirely one of the division of created wealth; that mere economic justice could not compensate for the loss of interest in work, which is the inalienable right of everyone, equally with the privilege of working and the right to earn a living. On the other hand it is impossible, impractical and undesirable that we should discard machinery. The machine is simply a tool, another means of expression developed through man's ingenuity. But we must develop some method of education which will bring workers more in sympathy with the machine, and increase their understanding of its significance. My own belief is that during the next few years in America, there will be a great increase in professional and rational craftsmanship; and that the great mill organizations, built so largely as ventures in finances, will split up into smaller units where personal direction and contact between executives and workers will be closer. So long as the problem remains an economic one, the contest between the selfishness of two groups, there can be but one answer; the strongest group will eventually win and in the long run labor will be the more powerful.

We have two forms or organizations of labor now existing in the textile field, one known as the craft union where each process of manufacture has its separate union organization. This type of textile union is a part of the American Federation of Labor and is strongly represented in certain sections of New England and in the South. There is another type of union, which includes in one single organization all workers in textile mills and is known as the Amalgamated Textile Workers of America, or the One Big Union Idea. This is largely a fighting organization, which comes into existence more strongly in times of controversy than in times of peace.

If the entire industrial North, East and South should become organized on this basis, all that labor would have to do to either wreck the industry completely or to have their demands met to the last iota would be to refuse to do anything; to stand still. Capital rules only through division of counsel in labor. It is necessary, therefore, that sympathetic accord and understanding and a common interest and satisfaction in work should be established.

There have been in the past a few prophets who have written books, and not a few individuals who wrote books they intended to be prophetic. Among the latter are to be numbered those who predicted a generation ago that staple cotton mills could not prosper in the South. Perhaps there may be others of the same high mentality who will now sign the death warrants for all cotton mills in the East.

Without aspiring to the full dignity of a true prophet while earnestly desiring to avoid the other rôle, there are a few safe generalities upon which I might venture.

The future of cotton mills is not a sectional question in any sense of the word. Cotton mills will survive or perish in either section only as they serve with intelligence and genius the public. Our public is fabric pampered, accustomed to the products of the world's finest looms and no way has yet been devised to force or cajole them into any other attitude. No great class of our citizenship fortunately is compelled by poverty to buy merely merchandise. Fabrics must appeal to them in some way to enjoy their patronage. Fashion spreads from our great cities as fast as the mail can carry the fashion news, as swiftly as merchandise can be shipped to points of distribution, and fashion is the modern, commercial term for beauty and taste.

Mills able and willing to meet these shifting requirements will prosper. The others will continue to squabble over a rapidly diminishing market. There is no question that the taste of America is already far beyond the average of industrial expression and the end is not yet in sight, nor is the public interested in the general economic conditions of the mills, nor the relationship these mills bear to the general position of America in the markets of the world.

The export trade, that once offered a periodical outlet for the accumulated staple merchandise of low grade, may never again be consistently relied upon for the simple reason that the manufacturers of textile machinery are rapidly supplying loomage and spindlage to these countries to enable them to make their own cheaper merchandise. These machines are so marvelously perfect and automatic that staple constructions can now be manufactured by almost any class of labor willing to stand and watch the machines run for ten or twelve hours a day at small wages and under expert direction. Local tariffs will take care of the slight differences in cost.

Machinery exports into Japan, China, India, Brazil, the Argentine, and other once great markets for machine-made staple cottons would tell an interesting story, if all the facts might be gathered and analyzed. Russia, for the time at least, imports no fabrics at all and is intent on rehabilitating her old machinery. Here is an illustration, directly in point, of how few staples are actually needed in any regions.

Before the War, Russia had over seven million cotton spindles and imported as well many millions of rubles worth of cloth. The Revolution and the loss of Finland cut her spindlage to three million spindles, archaic in type and not in particularly good running order. At the same time, with this inadequate equip-

Conclusion

ment and with outside supplies cut off, she has been able to satisfy her most pressing requirements and to successfully exclude all foreign cloth.

Russia will continue to buy her cotton from us, mill findings and machinery, and perhaps brains to run her mills for her until she can train non-political foremen; she will buy nothing else, until her economic system breaks down or changes.

India may follow suit and China may become easily equally independent, alike of our raw cotton and cotton goods and buy only machinery from us.

In Brazil there are over sixty active mills, fully equipped with modern machinery and finishing plants, and the Argentine is only a little behind. Mexico has a large population of sufficient intelligence to work in mills and has as well the possibility under settled conditions of raising large supplies of cotton. In Mexico today there are already several well equipped mill organizations, fully as efficient as any we have in this country making similar grades of merchandise.

The only kind of cotton goods, therefore, that we will be able to make or sell successfully either at home or abroad, will be the higher qualities of merchandise, requiring the best and newest machinery, the highest type of superintendence, the best trained help and the guidance of mill treasurers who realize the vital importance of design and fashion. This means in many instances a great change in the present organization of our mills, a greater degree of flexibility and the building up of healthy and contented labor communities of skilled workers. It means that each organization will have as a vital element a department of research and experiment in design.

Such mills, wherever founded, will succeed and their

success will be of economic and social value to this devoted and long suffering land. The fact that automatic machinery is one of the great conquests of modern civilization, does not change the fact that people desire merchandise for its æsthetic rather than its economic value; and the attitude of the world towards beauty has not changed merely at the behest of mentally stagnant mill treasurers who have followed out the ideas developed by a small group of brilliant mechanics in England, in the latter part of the Eighteenth Century.

If this be prophecy, then I stand indicted. For as far as I am personally concerned, the other kind of cotton mills may go to any country sufficiently deluded or benighted to desire them. They have no place in the economic or social scheme of America.

And so at the end of the story of cotton, we come again to that ancient and eternal desire for beauty, which launched ten thousand keels in quest of loveliness.

When we have finally mastered the true significance of the machine and raised it to its highest potential power, we may find its standards of production will bear comparison with the achievements of the golden yesterdays of the craft ages.

BIBLIOGRAPHY

ALI, AMEER:—"History of the Saracens."
APPLETON, NATHAN:—"Introduction of the Power Loom and Origin of Lowell." Lowell, Mass., 1858.
BAINES, EDWARD, JR.:—"The History of Cotton Manufacture in Great Britain."
BAKER, G. P.:—"Calico Painting and Printing in the East Indies in the Seventeenth and Eighteenth Centuries." Beautifully illustrated with color plates with interesting text. London, 1921.
BAND, H. R.:—"Indian Dyeing and Block-Printing."
BANERJEI, N. N.:—"The Cotton Fabrics of Bengal." *The Journal of Indian Art and Industry.* Vol. VII. No. 68. October 1899.
BARKER, A. F.:—"An Introduction to the Study of Textile Design." London, 1903.
BIGWOOD, GEORGE:—"Cotton."
BOWMAN, F. H.:—"Structure of the Cotton Fibre." London, 1908.
BROOKS, EUGENE CLYDE:—(Professor of Education in Trinity College, Durham, N. C.). "The Story of Cotton and the Development of the Cotton States."
CAREY, M.:—"A View of the Ruinous Consequences of a Dependence on Foreign Markets for Sale of the Great Staples of this Nation, Flour, Cotton and Tobacco." (1820).
CHATTERJIE, A. C.:—"Notes on the Industries of the United Provinces." Allahabad, 1908.
CHRISTIE, MRS. ARCHIBALD H.:—"Embroidery and Tapestry Weaving." London, 1906.
COOMARASWAMY, A. K.:—"The Arts and Crafts of India and Ceylon." London, 1913.
CRAWFORD, M. D. C.:—"Peruvian Textiles." (Anthropological Papers of the American Museum of Natural History.) 1915.

Bibliography

CRAWFORD, M. D. C.:—"Peruvian Fabrics." (Anthropological Papers of the American Museum of Natural History.) 1916.

DONNELL, E. J.:—"Chronological and Statistical History of Cotton." Published by the author, New York, 1872.

ENTHOVEN, R. E.:—"Cotton Fabrics of the Bombay Presidency." *The Journal of Indian Art and Industry*, Vol. X. No. 82. April, 1903.

FALKE, OTTO VON:—"Kunstgeschichte der Seidenweberei." Berlin, 1921.

FREZIER, MONSIEUR:—"Frezier's Voyage to the South Sea." London, 1717.

GARCILASSO DE LA VEGA:—"Royal Commentaries of Peru." E. Rycant. London, 1688.

GRIFFITHS, JOHN:—"The Paintings in the Buddhist Cave-Temples of Ajanta." London, 1897.

GUARANTY TRUST COMPANY OF NEW YORK, 1919:—"The Fabric of Civilization: A Short Survey of the Cotton Industry in the United States."

HADAWAY, W. S.:—"Cotton Painting and Printing in the Madras Presidency." Madras, 1917. The best general work on the cotton printing industry in India with excellent illustrations, by the Superintendent of the School of Arts, Madras.

HADDON, ALFRED C.:—"Evolution in Art: as illustrated by the Life-Histories of Designs." New York, 1914.

D'HARCOURT, R. AND M.:—"Art Ornemental les Tissus Indiens du Vieux Pérou." Éditions Albert Morancé, 1924.

HENDLEY, P. H.:—"Asian Carpets." Many rug designs were originally cotton designs.

HOLMES, WILLIAM H.:—(a) "Textile Art in its Relation to the Development of Form and Ornament." (Sixth Annual Report, Bureau of American Ethnology, Washington, 1889.) (b) "Textile Fabrics of Ancient Peru." Washington, 1889.

HOOPER, LUTHER:—"Hand-Loom Weaving, Plain and Ornamental." London, 1910.

KARLIN, A. M.:—Article appearing in *"Der Konfectionaer."* September 16th, 1923.

KIPLING, J. L.:—"The Industries of the Punjab." *The Journal of Indian Art and Industry*. Vol. No. 23. July 1888.

LE COQ, A. VON:—"Chotscho." Berlin, 1913.

Bibliography

MACCURDY, GEORGE GRANT, PH.D.:—"A Study of Chiriquian Antiquities." A scientific description of the pottery arts of prehistoric Central America. Yale University Press, 1911.

MARSDEN, RICHARD:—"Cotton Spinnings: Its Development."

MATTHEWS, J. MERRITT:—"The Textile Fibres." New York, 1924.

MITCHELL, BROADUS, PH.D.:—"Rise of the Cotton Mills in the South." Johns Hopkins Press, 1921. A careful and economic history of the cotton industry in the South.

MUKHARJI, T. N.:—"Art Manufacture in India."

MURPHY, WILLIAM S.:—"The Textile Industries." Eight volumes. London, 1912. Technical history of textiles fully illustrated.

NYSTROM, PAUL H., PH.D.:—"Textiles." New York, 1919.

OPPEL, PROF. DR. A.:—"Die Baumwolle: Nach Geschichte, Anbau, Verarbeitung, und Handel, sowie nach ihrer Stellung im Volksleben und in der Staatswirtschaft." (Compiled by request of the Bremen Cotton Exchange.)

PERCIVAL, MAC IVAR:—"Chintz Book."

PERSOZ, J.:—"Traite Theorique et Pratique de l'Impression des Tissus." Paris, 1846, four volumes of text with Atlas. A practical treatise on the chemistry of dyes and the art of dyeing, valuable for the many specimens of actual dyed fabrics, chiefly cotton, with which the work is illustrated.

REISS, W., AND STUBEL, A.:—"The Necropolis of Ancon in Peru." A contribution to our knowledge of the culture and industries of the Empire of the Incas, being the results of excavations made on the spot. Translated by Prof. A. H. Keane. Three volumes. Berlin, 1880–1887.

SAMMAN, H. F.:—"The Cotton Fabrics of Assam. *The Journal of Indian Art and Industry.* Vol. X. No. 82. April 1903.

SCHMIDT, MAX:—"Über Altperuanische Gewebe." (Baessler Archiv, Band I. Leipzic and Berlin, 1911.)

SHORT, ERNEST HENRY:—"Man and Cotton."

SILBERRAD, CHAS. A.:—"Cotton Fabrics of the North-Western Provinces and Oudh." *The Journal of Indian Art and Industry.* Vol. X. No. 82. April 1903.

SPINDEN, HERBERT J., PH.D.:—"A Study of Maya Art." Peabody Museum of American Archaeology and Ethnology.

Bibliography

Harvard University, 1913. A carefully illustrated survey of the prehistoric arts of Central America.

SPINDEN, HERBERT J., PH.D.:—"Ancient Civilizations of Mexico and Central America." American Museum of Natural History, 1917. A popular discussion of the ancient arts of Mexico.

SQUIER, E. G.:—"Incidents of Travel and Explorations in the Land of the Incas." New York, 1877.

STEIN, M. AUREL:—"Ruins of Desert Cathay." Two volumes. London, 1912. A personal narrative of explorations in Central Asia and westernmost China, with numerous illustrations, color plates, panoramas and maps from original surveys.

THURSTON, EDGAR:—"The Cotton Fabric Industry of the Madras Presidency." *The Journal of Indian Art and Industry.* Vol. VII. No. 59. July, 1897.

TINGLEY, RICHARD HOADLEY:—"The Dramatic Story of Cotton." From the *Mentor* of August, 1923.

WALTON, PERRY:—"The Story of Textiles." Boston, 1912.

WARNER, SIR FRANK:—"The Silk Industry of the United Kingdom." Its Origin and Development. London.

WATSON, J. FORBES:—"The Textile Manufactures and the Costumes of the People of India." London, 1866. A catalogue and key to a collection of specimens of all the important textile manufactures, comprising 700 specimens, (in large part cotton) contained in eighteen large volumes of which twenty sets were prepared from the store of the Indian Museum, Calcutta. A set of this most important work can be consulted in the Library of the Victoria and Albert Museum, London. There is a similar collection in eight volumes with printed preface dated 1851, Berlin, collected by the brothers Schlagintwert, chiefly from Nepaul in the Museum of Ethnology in Munich.

WATSON, WILLIAM:—"Textile Design and Colour." London, 1912.

WATT, SIR GEORGE:—"The Wild and Cultivated Cotton Plants of the World." London, 1909.

WEEDEN, WILLIAM B.:—"The Art of Weaving: A Handmaid of Civilization." (Annual Report of the American Historical Association for 1902, Vol. I, pages 191–210. Washington, 1903.)

Bibliography

Wiener, Charles:—"Pérou et Bolivie, Récit de Voyage suivi d'Études Archaeologiques et Ethnographiques et de Notes sur l'Ecriture et les Langues des Populations Indiennes." Paris, 1880.

Wiener, Leo:—"Africa and the Discovery of America," volume 2. Philadelphia, 1922. Apart from theory contains valuable information on cotton in Africa and aboriginal America.

Woodbury, Levi:—(Secretary of the Treasury) "Tables." Published in 1836.

Zipser, Julius:—"Textile Raw Materials and their Conversion into Yarn." Edited by Charles Salter. London, 1901.

INDEX

A

Amalgamated Textile Workers of America, 228
American Federation of Labor, 220
American Museum of Natural History, 199
Anabaptists introduce printing in southern Germany, 83
Appleton, Nathan, 145, 146
Armada, Spanish, 85
Arkwright, Richard, 111, 112, 115, 116
Art Alliance of America, 200
Artificial Silk, 189
    The American Tubize Co., 191
    The American Viscose Co., 191
    Andemars, Swedish chemists 1858, 190
    Celanese, 192
    Cross, Bevan and Beadle, 1892, 190
    Cotton cellulose, 178
    Count Hillaire de Chardonnet, 1889, 190
    The Industrial Fiber Co., 1732, 191
    Reaumur, French Chemist, 1734, 190
Artists, Lists, 202
Atlanta Exposition, 169
Ayres & Co., L. S., 205

B

Baker, G. P., 75
Ball, Katherine W., 20
Barbosa, Obvarado, 74
Basset, Wm. R., 193
Baumgarten, Fannie, 203
Bell, Thomas, 119, 145
Bennett, Thomas, Jr., 152
Beverly, Mass. 134
Bihr, Alphonse, 202
Blas, Gus, Co., 205
Blum, Albert, 198, 201
Boll Weevil, 223
Bonwit Teller & Co., 205
Bow, The Missile, 17
Bowers, Mrs. Isaac, 158
Brava, Island of, 155
Brooklyn Museum, 203
Budd, F. W., 202
Burt, W. G., 201

C

Cabot, Mr., 134
Calicoes; great demand for, 96
Carlyle, Thomas, 121
Carr, Maria C., 202
Cartwright, Edmund, 118, 119, 145
Chadwick, George B., 201
Charlotte, N. C., 166
Cheney, Charles, 202
Chinese Explorers, 64
Clendenning, Mary Louise, 202
Cochineal, 85
Cœurdoux, Father, 75
Colonial Textile Arts, 127
Color, The beginning of color application, 21
Columbus finds cotton in Brahma Islands, 30
Compotus, Earl of Derby, 1381, 90
Compotus, Bolton Abbey. First mention of cotton in England, 90
Cone, Julie, 202
Contributors, List of, 201
Converters, 154
Cotton:
    Arabic names for cotton, 63
    Brahmans and cotton, 66
    The cotton buyer, 180
    Classical names of cotton, 62
    Cotton exchanges, 217
    Cotton in Europe, 81
    Cotton shipped to England, 1764, from Colonies, 129
    Cotton Machinery Exposition in Greenville, S. C., 204
    Cotton in 1140 in Genoa, 82

Index

Cotton—*Continued*
  Germany receives cotton from Brazil, 1570, 125
  Grades of Cotton, 181, 214
  Law in 1721 regarding use of cotton, 97
  Law forbidding sale of cotton, 1700 and 1712, 96
  Cotton seeds for Colonists, 126
  Cotton shipments from 1791–1911, 139
  Cotton in Spain, 81
  Cotton in Ulm in 1320, 82
  Upland cotton, 130
Crawford, M. D. C., 202
Crompton, George, son of inventor, 100
Crompton, Samuel, 114, 116, 117, 121
Crompton, Samuel invents spinning mule, 1779, 115
Culin, Stewart, 201, 204
Culin, Stewart, The Missile Bow, 17

D

Dacca muslins, 23
Dacca muslins and Indo-Greco statuary, 66
Dacca muslin names, 67
Dallimore, Alice L., 203
Daly, Helen S., 202
Davidson, Henry P., 201
Davidson Bros., 206
Decoration, 21
Defoe, 98
Denecke Co., The, 206
Discovery, Era of, 74
  Vasco da Gama passage to India, 74
Doughty, Emma W., 203
Drake, Sir Francis, and the *St. Phillip*, 85
Drew, Louise, 203
Duffy, Y. Constance, 203
Dupont Co., 191
Dyes:
  German Cartel in Dyes, 186
  Dye Industry, 186
  Dye Industry in U. S. result of war, 186
  Dyes and explosives, 187
  Vat Dyes, 189
Dyeing:
  Resist dyeing in pre-historic Peru, 58
  Pliny describes Mordant dyeing, 65
  Distribution of Resist Dyeing, 76

E

Eakin, L. R., 205
Emporium Mercantile Co., 206
England:
  Cotton used as Candlewicks, 12th Century, 4
  English difficulty with the Dutch, 94

F

Fairchild, E. W., 197
Field, L. H., Co., 206
First International Cotton Conference, 220
Fleary, Adrien, 202
Franklin, Benjamin, 127, 128
Folsom, Frederick C., 201
Frederick & Nelson, 206
Frietsch, Cholly, 203
Fulton, Francis, 202

G

da Gama, Vasco, 72, 85
Garrick, David, 101
Gin, the Cotton, and Slavery, 139, 163
Gladding, Hope, 202
Gourney, John describes cotton trade with India, 1614, 93
Gowing, Charles, 20
Gregg, William, 164
Greeks, knew of cotton and cotton technique before Christian Era, 6
Green, Mrs. Nathanial, 138
Greenville, S. C., 166
Grinnell, Joseph, 152

H

Hall I' Th' Wood, 114
Hammond, Senator of S. C., 170, 141
Hanson, E. Irving, 201
Hargreaves, James, 134, 115
Harned & Van Maur, Inc., 206
Hecht, Zolton, 202
Heinke, A. J., 203
Hentschel, M. E., 202
Herter, Albert, 201
Hochschild Kohn & Co., 205
Hoover Committee Elimination of Waste in Industry, 193
Hopedale, Massachusetts, 122
Hopi Cottons of Ceremony, 37
Hovey Co., C. F., 205
Howland, The, Dry Goods Co., 205
Hughs, Bess B., 203
Hurd, Alice M., 202

Index

I

India:
Methods of decorating Indian cotton, 75
Indian costumes, 71
Seventeenth Century Indian Cotton, 70
Indigo, Herodotus mentions, 64

J

Jackson, Patrick T., 146
Jacquard, Jean Marie, 119
Java, received cotton from India, 3rd to 5th Century A.D., 7
Jeffersonian embargo, 145
Jeffrey, Dr., 19
Jelleff, Frank R., Inc., 205

K

Karasz, Ilonka, 203
Kaufman & Baer Co., 205
Kaufmann Strauss Co., 205
Kennedy, John, 116
Kent, Henry W., 200, 201
Kinloch, Andrew, 119
Kipling's *Naulahka*, 70
Kramer, Conrad, 203
Ktesias mentions cotton, 400 B.C., 64

L

"Lagoda," 150
La Grange, Ga., 166
La Salle & Koch Co., 205
Lawrence, Amos & Abbott, 149
Lawrence, Lillian, 202
Long, Florence, 202
Leverholm, Lord, 114
Lewis, The A. T., & Son Dry Goods Co., 206
Lowd, Edna B., 203
Lowell, Mass., 148
Lowell, Francis C., 145, 146, 147, 148, 149
Looms:
Two basic types of looms, 25
The loom and Cloth Making, 24
Description of Cotton Loom, 25
Distribution of Cotton Looms, 26
Draper Loom, 121
Fourteenth Century Loom in England, 107
M. de Gennes makes drawing of power loom, 1678, 106
M. de Gennes, 118
Looms of Haida Tribes in Alaska, 26
Introduction of Indian Loom in England and Europe, 28
Age of Indian Loom, 66
Indian forms of Cotton Looms, 28
John Kay of Bury and fly shuttle, 108
Francis C. Lowell and the Power Loom, 145
Northrop Loom, 122
Parallel distribution of bow and single barred loom, 34, 35
Penelope's Loom, 26
Peruvian Loom, 56
Prehistoric and Modern Looms in New World, 33
Warp Weighted loom from Swiss Lake Cultures, 25

M

MacLaren, J. W., 201
Manu, Cotton in Statutes of, 65
Marshal, Mary, 202
Mayer, Edward L., Costumer, 201
McDowell, James, 182, 184
Megasthenes, 300 B.C., mentioned flowered muslins, 65
Meier & Frank Co., 206
Memphis, 220
Metropolitan Museum of Art, 199
Meyer, Max, 201
Migel, J. A., 201
Mijer, Pieter, 202
Miller & Paine, 206
Mitchell, Broadus, *The Rise of Cotton Mills in the South*, 163
Mills:
Acushnet Mills, 153
Amoskeage, 1831, 150
Bennett Mfg. Co., 153
The Booth Mill, 149
The Boston Mfg. Co., 146, 148
Bristol Mfg. Co., 153
Butler Mills, 154
The Charleston Mfg. Co., 170
City Mfg. Co., 153
Columbia Spinning Co., 153
Cotton Mills in Haverhill, Salem, Nantucket and Exeter, 150
Dartmouth Mfg. Co., 154
Grinnell Mfg. Co., 153
Gosnold Mills Co., 154
The Great Falls Mfg. Co., 1823, 150
The Hamilton Co., 149
Hathaway Mfg. Co., 153
Howland Mfg. Co., 153

Index

Mills—*Continued*
  Kilburn Mills, 154
  Laconia Mills—1845, 150
  The Lawrence Co., 149
  Manomet Mills, 154
  The Massachusetts Mill, 149
  Merrimack Mfg. Co., 148, 149
  Naumkeag Steam Cotton Mill, 1847, 150
  New Bedford Mfg. Co., 153
  Nonquit Spinning Co., 154
  Pacific Mfg. Co., 1854, 150
  Page Mfg. Co., 154
  Pepperhill Mills, 1850, 150
  Pierce Mfg. Co., 153
  Potoomska Mills, 153
  Quansett Spinning Co., 154
  Rotch Spinning Co., 154
  Slater Mills, 144 147,
  Soule Mills, 154
  The Suffolk Mill, 149
  Taber Mill, 154
  The Tremont Mill, 149
  Type of fabric made in Waltham, 147
  Wamsutta Mills, 1847, 152–153
  Whitman Mills, 154
Mogul Empire, Fall—1368, 73
Mohammedan Conquests and Cotton, 68
Mohammedan-Pre Cotton designs, 69
Molière's *Le Bourgois Gentilhomme*, 86
Moody, Paul, 146 148,
Moors, The, Abdrahaman III, 81
Morehouse Martens Co., The, 205
Morey, Bertha, 202
Morris, Wm., 121

N

National Aniline & Chemical Co., 80
National Association of Cotton Manufacturers, 205
New Bedford, 151
Neiman-Marcus Co., 205
New World:
  Antiquity of cotton in New World, 32
  Asia, culture home of America, 34
  Asiatic intrusion, 49
  Cotton blankets in tribute rôle of Montezuma, 40
  Cotton from Chichen Itza, 41
  Discovery of cotton in New World great surprise, 8
  Law of the Indies, 41
  Maya culture, 32
  Native dyes, 44, 45
  Parallel distribution of bow and single barred loom, 34
  Pre-historic cotton culture, 31
  Pre-historic cotton map, 31
  Spanish influence on native design, 42
Neyland, Harry, 201

O

Oberkampf, Christopher Phillip, Jouy, 87
Owen, Robert, 121

P

Peel, Sir Robert, 149
Percival, Spencer, 117
Perry, Dwight, 152
Peru:
  Burial customs, 52
  Peruvian colors, 50
  Peruvian design, 60
  Types of Peruvian fabrics, 56
  Incas, 3
  Pre-Inca, 3
  Cottons of Pre-Inca Peru, 53
  Pre-Inca culture, 50
  Preparation of fiber, 53
  Resist-dyeing in Pre-historic Peru, 58
Petremont, Clarice, 203
Pintado, 74
Pikula, Anna, 202
Plague among dyers on Coromandel Coast, 95
Pliny, 75
Polo, Marco:
  States cotton was only known in Fokien, 7
  Discovered indigo in 1300, 84
  Chintzes of Masuliputam, 73
Pogue, H. & S. Co., 205
Poor, Marion, 203
Portuguese, The, 155
Prendergast, Charles, 201
Printing:
  Thomas Bell invents copper rollers for printing in 1770, 103
  Charles Taylor and Thomas Walker, 103
  List of printing plants in England and Europe, 84
  Manchester Act in 1736 permitting manufacture and sale of British calicoes, 101
  First British printing plant established in Richmond in 1690, 95
Purdy, F. W., 201

Index

R

Reeves, Ruth, 203
Reiss, F. Wienold, 204
Rodier, Paul, 197
Roe, Sir Thomas, 70
Roessel, Carl, 202
Roosevelt dam, Salt River Valley, 180
Ruskin, John, 121
Ryther, Martha, 202

S

Salem, Massachusetts, 128
San Blas Indians design, 44
Scythian Sheep—Cotton in the Middle Ages in Europe was supposed to be the wool of a vegetable sheep, 5, 63
Silk in colonies, 126
The rise of the Silk Industry, 160
Simonson, Grace H., 202
Slater, Samuel, goes into partnership with Silas Brown and builds first yarn mill in 1793 in Pawtucket, 135-136
Slaughter, Hazel Burnham, 202
Smith, Bertha, 202
Smith, Rachel, 202
Smiths, early mechanical efforts of the, 135
Southwest, Oldest Cotton fabrics from, 37
  Antiquity of Cotton, 38
Spanish Colonists, 125
Spinden, Dr. Herbert J., 32
Spinning: 54
  Two methods of spinning, 23
  Spinning wheel, 24
  James Hargreaves Spinning Jenny, 110
  Leonardo da Vinci invents spinning flyer, 106
  Old Saxony Wheel, 127
  Louis Paul Carder, 110
  John Wyatt Spinning Rollers, 110
  First spinning mill, in Philadelphia, 134
  Ring frame, 122
Cotton statistics:
  From 1815-1859, 140
       1859-1860, 141
       1864-1865, 142
Stein, Sir Aurel discovers cotton fabrics in Gobi Desert, 2nd to 5th Century, 6
Stix, Baer & Fuller Dry Goods Co., 205
Stores, List, 205

Strutt, Jedediah, 112
Sumene, Mlle. Durant de, 202

T

Tannahill, Mary, 202
Tash, John, 93
Textile machinery, 166
Textile workers migrate from Continent to England, 91
Thalheimer Bros., 205
Thompson, J. H., 201
Tilden, Alice F., 202
Trade:
  Arab Merchants, 69
  The British East India Co., 86, 125
  The Dutch East India Co., 86
  Dutch opens trade in cotton with Japan in 17th Century, 7
  The French East India Co., 85
  Development of Trade in the Orient, 85
  Trade between the Orient and Europe, 72
  Egypt, no cotton in commerce with India nor silk with China, 7
  Cotton trade in Pre-Spanish America, 39
  Mr. Sedgwick presents British calico to Princess of Wales, 102
  Rise of Turks cuts off Europe's trade, 73
  The Weavers' True Case, 101
  Moors introduced cotton into Spain, 9th and 10th Century, 4

V

Vogel, Milton, 201

W

Waltham, Mass., 146
Walton, Laura E., 202
War of 1812, 145, 147
Ward, B. C., & Co., 158
Washington, George, 134
Water Sheep; Chinese explorers early Christian Era mention "Water Sheep" perhaps referring to cotton, 8
Watt, James, steam-engine, 120
Waugh, Coulton, 203
Wearne, Harry, 201
Whaling industry in New Bedford, 151
Whitman, David, 152
Whitney, Eli, 119, 142
  Whitney saw tooth Gin, 1793, 138, 163

Whitney-MacGregor Co., 206
Wilson, Ruth J., 202
Wissler, Dr. Clark, 199
Witters, Nell, 202
Women's Wear, 199
Wool merchants protest use of cotton, 1621, 92
Wyatt, Jno., 115

Y

Yarns, 55
Yetters, 206

Z

Zorach, Margurite, 202